CAMBRIDGE LIBRARY COLLECTION

Books of enduring scholarly value

Life Sciences

Until the nineteenth century, the various subjects now known as the life sciences were regarded either as arcane studies which had little impact on ordinary daily life, or as a genteel hobby for the leisured classes. The increasing academic rigour and systematisation brought to the study of botany, zoology and other disciplines, and their adoption in university curricula, are reflected in the books reissued in this series.

Observations in Natural History

Although devoted to his parish, Leonard Jenyns (1800–93) combined his clerical duties with keen research into natural history. Notably, he was offered the place on the *Beagle* that later went to Charles Darwin. His numerous works include *A Manual of British Vertebrate Animals* (1835) and *Observations in Meteorology* (1858), both of which are reissued in this series. First published in 1846, the present work was originally intended as a companion volume to Gilbert White's acclaimed *Natural History and Antiquities of Selborne* (1789), which Jenyns had copied out as a student at Eton. The product of two decades of meticulous observation of Jenyns' surroundings in eastern England, the text includes journal entries with careful records on a wide variety of wildlife, including quadrupeds, birds, reptiles, fish, insects and molluscs. Also featuring a detailed calendar of periodic phenomena, this work illuminates the rhythms and quirks of the natural world in England.

Cambridge University Press has long been a pioneer in the reissuing of out-of-print titles from its own backlist, producing digital reprints of books that are still sought after by scholars and students but could not be reprinted economically using traditional technology. The Cambridge Library Collection extends this activity to a wider range of books which are still of importance to researchers and professionals, either for the source material they contain, or as landmarks in the history of their academic discipline.

Drawing from the world-renowned collections in the Cambridge University Library and other partner libraries, and guided by the advice of experts in each subject area, Cambridge University Press is using state-of-the-art scanning machines in its own Printing House to capture the content of each book selected for inclusion. The files are processed to give a consistently clear, crisp image, and the books finished to the high quality standard for which the Press is recognised around the world. The latest print-on-demand technology ensures that the books will remain available indefinitely, and that orders for single or multiple copies can quickly be supplied.

The Cambridge Library Collection brings back to life books of enduring scholarly value (including out-of-copyright works originally issued by other publishers) across a wide range of disciplines in the humanities and social sciences and in science and technology.

Observations in Natural History

*With an Introduction on Habits of Observing,
as Connected with the Study of that Science*

Leonard Jenyns

CAMBRIDGE
UNIVERSITY PRESS

University Printing House, Cambridge, CB2 8BS, United Kingdom

Cambridge University Press is part of the University of Cambridge.
It furthers the University's mission by disseminating knowledge in the pursuit of
education, learning and research at the highest international levels of excellence.

www.cambridge.org
Information on this title: www.cambridge.org/9781108069861

© in this compilation Cambridge University Press 2014

This edition first published 1846
This digitally printed version 2014

ISBN 978-1-108-06986-1 Paperback

OBSERVATIONS

IN

NATURAL HISTORY.

OBSERVATIONS

IN

NATURAL HISTORY:

WITH AN INTRODUCTION ON

HABITS OF OBSERVING,

AS CONNECTED WITH THE STUDY OF THAT SCIENCE.

ALSO A

CALENDAR OF PERIODIC PHENOMENA

IN

NATURAL HISTORY;

WITH REMARKS ON THE IMPORTANCE OF SUCH REGISTERS.

By THE REV. LEONARD JENYNS, MA., F.L.S., ETC.,

VICAR OF SWAFFHAM BULBECK, CAMBRIDGESHIRE.

Multa enim in modo rei, et circumstantiis ejus, nova sunt, quæ in genere ipso nova non sunt: qui autem ad observandum adjicet animum, ei, etiam in rebus quæ vulgares videntur, multa observatu digna occurrunt.—BACON.

LONDON:

JOHN VAN VOORST, PATERNOSTER ROW.

M.DCCC.XLVI.

LONDON:
Printed by S. & J. BENTLEY, WILSON, and FLEY,
Bangor House, Shoe Lane.

PREFACE.

THE author of the present work, when engaged a few years back in preparing notes for a new edition of White's Natural History of Selborne, soon found a larger stock of matter collected upon his hands than it was thought desirable to use on that occasion. This led to the idea of embodying a considerable proportion of this matter in a separate work, such as is now offered to the public. It further appeared to him that a work of this kind might take the place of the " Naturalist's Calendar " and " Observations in various branches of Natural History," extracted from White's papers after his death, and first given to the world in a separate volume, by Dr. Aikin; both which have generally been incorporated with the later editions of White's work, but were not included in the edition spoken of above. This omission was in consequence of an opinion, that the " Calendar," though useful to a certain extent, wanted that method and exactness which alone can give any permanent value to such Registers; while the " Observations " contained little of importance beyond what might be found in the Natural History of Selborne itself, or what, in the case of the above edition of that work, it was easy to transfer to it in the form of notes.

The present volume, therefore, may be considered as an original work, offered in lieu of a reprint of the old one by Dr. Aikin, though for the most part similar to this last in plan and arrangement. The "Observations," which occupy a large proportion of it, are of a miscellaneous kind, and relate to various branches of Natural History; but they are all arranged under independent heads, and the animals to which they refer are classified according to their respective places in the system: by this means they will admit of being readily referred to as occasion may require. Of these observations a few have been contributed to the author by some of his friends, who have kindly interested themselves in the subject, and to whom he here begs to express his thanks; but the great majority of them have been made by himself, and principally in the immediate neighbourhood of his own residence. At the same time they have been made at very different times, over a considerable number of years, as the dates which are annexed to many of them sufficiently shew; some going as far back as the day, when he was only just commencing observer, and yet warm with the ardour inspired by a first perusal of White's own work. Indeed, not unfrequently the observations relate to matters, which the author had been led to notice from what White himself had previously observed; or they are accompanied by remarks suggested entirely by what the latter had written upon the same subjects. This will account for the frequent allusions to White,

as well as for a certain degree of similarity, which
perhaps may be traced in some parts of the book
to passages which occur in the Natural History of
Selborne. It is not thought that this last circumstance will render the public less favourable to the
undertaking; or that a work like the present will be
the less welcomed from the fact of several other
works, equally indebted for their origin to that of
White's, having already appeared at different times.
The works alluded to can scarcely be said to be
altogether upon the same plan as this one; and if
they had been, works of this kind can hardly be multiplied too much, so long as the observations they
contain are trustworthy and original.

Neither is the value of such works necessarily
limited by the degree of importance which may be
attached to these observations. The author is aware
that many of the observations here brought forward,
have little in them that is either likely to attract the
notice of the experienced naturalist, or calculated
materially to promote the advancement of any department of the science of which they treat. Nevertheless, he conceives, that they may not be, on the
whole, without interest; or unacceptable to such
young persons as are likely to be readers of the
Natural History of Selborne, to which work the
present may be considered as a companion. He is
disposed to hope that this volume may also have its
influence in tending to increase the number of
observers; that it may help to put those, who are
much abroad, or who are otherwise favourably circumstanced in this respect, into the way of collecting,

and entering in a daily journal, such little facts as may offer themselves to their notice, and which, under the guidance here afforded them, they may deem worthy of record.

It is especially with a view to this end, that the author has treated in a General Introduction to the work, of the value and advantage of those *habits of observing*, which are indispensable to the progress of all sciences, but especially of the science of Natural History, and which, at the same time, are the sources of so much delight and amusement to all true lovers of Nature. He need not here enter into details connected with this subject, which the reader will find in that Introduction stated at length. He may simply mention that, in addition to the remarks he has there offered on this head, he has drawn up a few rules, by attention to which, young naturalists may acquire the habit of *observing correctly;* from which also they may derive a few hints for carrying on their observations upon such a plan as would materially enhance the value of what they observe, as well as tend more surely to recompense them for the time and labour they had bestowed on such a purpose. For some of these hints, or for trains of thinking that have led up to them, the author himself has been indebted to certain parts of the admirable " Discourse on the study of Natural Philosophy," by Sir John Herschel.

With respect to the " Calendar," which forms the concluding portion of the present work, enough is stated in the Introduction to that particular part, to render any further explanation of it unnecessary.

It may serve in some respects to take the place of White's : but it is rather offered as a specimen of the kind of Calendar, which it would be desirable to see kept in different localities, in order to determine (what science at the present day principally requires), the *mean* date of occurrence of the several periodic phenomena observed in such localities. With this view, as well as for the purpose of shewing the coincidences more clearly, the plan of it is somewhat different from, and, as the author is inclined to hope, an improvement upon, that of the one compiled from White's Manuscripts by Aikin. The Calendar itself is, confessedly, very imperfect ; though, in respect of a considerable number of the entries contained in it, it may be regarded as a near approximation to the truth. Strictly speaking, also, it has no immediate reference to any other district, than that in which the author is resident, and in which the registered phenomena were severally observed : nevertheless, it may serve in a general way for all places in this country on or near the same parallel of latitude. In order to assist observers in comparing the dates of the periodic phenomena which they may notice in their own neighbourhood with those here given, as well as to facilitate the inquiries of persons generally, who may wish to know when any particular phenomenon takes place, there is annexed to the whole an Alphabetical Index having a reference to the *mean* date of occurrence in each particular case : any information that may be desired beyond this must be sought for in the Calendar itself. The author does not think it necessary to speak here of the utility of such Registers, as this point will be found amply discussed in the Introduction above alluded to.

In conclusion, it may be added, that though all the "Observations" which form the main part of this work are original, yet from the circumstance of their having been mostly made many years back, it has happened that similar ones, in some instances, have been since made by other persons. Where this has been the case, the observation has still been admitted; and a reference has been given in a note to those works (so far as they are known to the author) in which the observation already occurs, or in which there is anything connected with the particular fact to which it relates. The author has also transferred to the notes all the scientific names of the animals spoken of, (except where they have no English one generally received); as well as any matter of a more strictly scientific nature, which it was thought desirable to introduce, in further illustration of the particular subject under notice, but which was considered as not likely to be interesting to the general reader.

It was intended that this work, which was announced as in preparation a considerable time back, should have appeared before now, but the author found so much more labour in the arrangement of the Calendar than he at first anticipated, that his professional duties did not allow him the requisite leisure for getting the whole ready for publication sooner.

SWAFFHAM BULBECK,
July 6, 1846.

TABLE OF CONTENTS.

xii CONTENTS.

CONTENTS.

XV

PAGE.

INTRODUCTION

ON

HABITS OF OBSERVING,

AS CONNECTED WITH THE STUDY OF

NATURAL HISTORY.

Naturæ divitiæ planè sunt inexhaustæ, nec cuiquam post mille secula nato deerit quod scrutetur, et in quo se cum laude exerceat.—Raii *Epist.*

> Nature, enchanting Nature, in whose form
> And lineaments divine I trace a hand
> That errs not, and find raptures still renew'd,
> Is free to all men —— universal prize.
> Strange that so fair a creature should yet want
> Admirers, and be destin'd to divide
> With meaner objects ev'n the few she finds !
>
> Cowper.

B

INTRODUCTION.

———

(1.) NATURAL HISTORY has been said by some to be a study of facts; by others, a science of observation. Each of these statements is to a certain extent true : it is only by observation that we can acquire a knowledge of the facts upon which all ulterior views must be based. But neither does this, nor any other science, rest satisfied with the bare recording of isolated and independent phenomena. It seeks to classify these phenomena, and to comprise them within certain general principles, established by inductive reasoning, to the influence of which they may all ultimately be referred. Doubtless this is taking an extensive view of the subject, and, when we contemplate the immensity of nature, entering upon a field which it may require years to travel over, and which we may never be able to measure in its full dimensions. Yet that there are such principles as we allude to, is next to certain. Looking only to the analogy of other sciences, we might predicate their existence ; when, however, we regard further the frequent attempts which have been made of late years to discover and fix them, and the

B 2

present aspect of the science compared with what
it exhibited a century or two back, we can hardly
entertain a doubt on this point. And though we
may never attain to a complete knowledge of them,
they form as it were the main aim and object of the
science, to which the labours of the scientific naturalist are ever directed, and to which at least he
makes a nearer advance, the more he investigates
the relations existing amongst the various matters
that present themselves to his notice.

(2.) What has been said may seem, at first hearing, to discourage the labours of a certain class of
observers, who take the greatest delight in watching nature, but who can never hope to undertake
any such investigations as those above alluded to.
It may be thought to subject them to the charge of
resting satisfied with first steps, without caring to
advance in the true path of scientific research.
This, however, is far from what is intended ; and
it is rather in especial reference to these persons
that the remarks we are about to offer will be
brought together. A man may never aim at being anything more than a mere observer, and yet employ his time usefully to others, as well as agreeably to himself. He may restrict himself to simply
noting and recording what falls under his own
*autopsia,** and unconsciously be laying the foundation of the most important generalizations. For
observation, though not itself the true end of the

* This word was particularly used by White, to signify the
observing things for oneself and with one's own eyes.

science of Natural History, is nevertheless a means
to that end; and, whatever principles we ultimately arrive at, it is only observation that can have
insured their correctness or permanence. Hence the
facts and observed phenomena collected by such
persons may be of much value to others, though the
observers themselves make no immediate use of
them. And it has not unfrequently happened,
that the profoundest naturalists, whilst engaged in
the higher departments of the science, have ex-
pressed themselves indebted to some retired observer for the knowledge of a fact which has proved
of the greatest importance to their views, and been
one of the main supports of the theory they were
seeking to establish.

(3.) Of course it is presumed that the observations we propose making in any department of
nature should be correct, which is all that is necessary to give them value and importance; and,
without this, they will have no value at all.
It is essential to premise this, because some persons, who are not habituated to observing, may fall
unintentionally into errors, and, being first deceived themselves, may afterwards mislead others.
Further on it will be our object to lay down a
few rules, and to suggest certain hints, that may
prove of service in this matter, and by attending
to which, young observers may be put upon their
guard, and withheld from needlessly sacrificing both
their time and their labour.

(4.) But further; it has happened in most sciences that the collecting of facts, and the deducing

from them any such important generalizations as
may lead to a comprehensive theory, have been
the work of different individuals. In sciences, indeed, which have made any considerable advance,
such division of labour becomes unavoidable. The
observer, therefore, need not be discouraged, because he is not possessed himself of those attain-
ments necessary for proceeding to the investigation
of such principles as his observations may assist in
establishing. He may leave this to others, and
content himself, if he will, with a more subordinate part. And in Natural History especially it is
almost necessarily the case that the observer and
the theorist should be in some measure separated.
For the facts here required are of such a kind as
cannot all be procured at will, or in any situation
we please. Many of them call for opportunities of
a very peculiar nature; and those who enjoy such
opportunities are not unfrequently, by that very
circumstance, shut out from making any application themselves of the knowledge they have acquired to the furtherance of the general interests
of the science.

(5.) That we may see this the more clearly, let
us stop to take a general view of the descriptions
of facts which are wanted by the naturalist to
enable him to proceed in his inquiries into the
general principles upon which Nature seems to
have based her system. For this purpose he must
have under his view all the different species and
varieties of animals with which this earth is peopled; and he ought to be able to inspect them, not

merely in the dead or preserved state, but in the
living or at least recently-killed. Without this he can
never closely investigate, or fully understand, many
parts of their structure, the knowledge of which
is sometimes essential for forming even the most
distant idea of their true affinities. Besides this,
there are many tribes of animals, especially among
the lower classes, which, from their delicate organization, are scarcely capable of being preserved at all.
To study these, therefore, and to obtain even the
most general view of their organization, it will be
necessary for him to resort to the spots where they
are found. But further; in addition to a knowledge
of structure, the naturalist requires a knowledge
of habits. The former, indeed, is always connected with, and more or less subordinate to, the latter. And though he may sometimes infer a particular habit from a given structure, yet such inference can only be the result of having first actually observed their co-existence in a large number of instances. After all, what is more varied
than the habits and instincts of different animals?
How multiplied are the resources of Nature in com-
passing her ends! How often do we find the same
object attained in as many different ways as there
are cases in which the object is sought! Again,
how strangely are both habits and structure sometimes modified by accidental circumstances, and by
conditions affecting certain particular localities! —
So that it is at once obvious, that, in order to become acquainted with all the phenomena of Natural History, a man must leave the retirement of

his study, and pursue Nature into her own haunts,
amid fields, waters, woods and mountains, (to say
nothing of travelling to foreign climes,) or else he
must have a large proportion of such facts collected to his hand by others, upon whose accuracy
he can depend.

(6.) This circumstance has given rise to a distinction between (as they have been sometimes called) *in-door* and *out-of-door* naturalists. Every one
who has read White's Selborne, is aware how its
estimable author uses this latter term in speaking
of himself, not professing to be anything more than
a close observer of nature, and leaving to others to
compare and classify the facts so obtained, and to
build up systems in their own closets at home.
Neither is this distinction a bad one. There is
a great deal, as regards the real advancement of
the science of Natural History, which can be done
only at home, where there is quiet and leisure, together with ready access to a well-stored library;
and there is a great deal likewise, as we have just
seen, that can be done only abroad. And it is ab-
surd for either of these two classes of naturalists
to throw contempt and censure upon the other, as
sometimes has been the case; seeing that they both
work together for the good of the science, and labour in a common cause, although in different ways.
The in-door naturalist cannot do without the out-
of-door; and the latter, one might suppose, would
never undervalue the inquiries of the former,
which tend to increase the importance of his own
researches.

(7.) Each of the above habits of life has its peculiar advantages; but the out-of-door naturalist, with whom alone we have to do in these pages, has this in his favour, that he is never at a loss for the means and opportunities of carrying on his favourite pursuit. He is, as it were, independent of circumstances; and he can scarcely be so situated as to be altogether shut out from observing nature, and adding to the store of facts which it is his amusement to collect and note down. All places offer something worthy of his attention, though certain spots may be more favourable in this respect than others. And here we shall make a remark which may be of service to young naturalists, viz. that they should avail themselves of the particular situation in which they may happen to be placed, for studying whatever that situation offers. We mean, that they should be guided by it in their choice of the particular department of Natural History to which they intend mainly to confine their researches. They should take what comes to hand, and bestow their chief time on what falls directly under their eye. They will often thus have it in their power to supply facts for the more scientific naturalist, which others, differently circumstanced, have no means of ascertaining. For almost every locality has its peculiarities; and we should endeavour rather to turn these to account, than to be on the constant search for what our neighbourhood is not calculated to afford. We like the remark made by White in one of his letters to Pennant, that if he had been by the sea-side he would have turned his attention to

Ichthyology.* Can there be a question as to whether
that observant naturalist, had he been so situated,
would not have given us just as entertaining and
instructive a volume, as that he has left us in his
Natural History of Selborne? It was not the exact
locality in which he was placed, but the inquisitive
turn of his mind, and the ardent zeal with which he
devoted himself to the study of Nature, that enabled
him to glean those materials, which have afforded so
much delight to the present generation. He himself observes in another passage of his book, "that
that district produces the greatest variety, which is
the most examined;"† and few, who reflect on the
many novelties constantly occurring in certain districts, where there happen to be more observers
than elsewhere, will be disposed to question the
general truth of this statement.

(8.) It is doubtless in a great measure owing to
the influence which White's Natural History of
Selborne has exercised on the present generation,
that the science has had so many followers in this
country of late years. It is not that his work
carries us any great way in unravelling the mysteries
of Nature, but it is the spirit which it breathes that
so strongly recommends it to our notice. He has
induced others to follow up the same sort of life
which had such charms for himself; and to him we
are indebted for many volumes besides his own, of
which the authors, by their own acknowledgment,
were first excited and trained to habits of observing,

* *Nat. Hist. of Selb.* Lett. xxi. to Pennant.
† *Id.* Lett. xx. to the same.

by the perusal of his work. He has taught us how much there is of interest to an intelligent and inquiring mind in the history of the meanest and most familiar of animals. Naturalists are not now, so much as formerly, content with the discovery of new species, nor desirous merely of enriching our museums with rare and costly specimens; but they seek to know the economy and mode of life of the different animals they meet with. Many of these they have traced with the greatest assiduity and patience through their several changes dependent upon age, as well as watched the modifications induced by varying circumstances on their particular habits. They have learnt to attach a value to little points in their history, which formerly would have been passed over as unimportant, or perhaps not observed at all. And it is greatly to be desired that all professed collectors would follow such a course, and note down whatever facts they can get hold of in illustration of the specimens they acquire. This practice would much enhance the value of their collections, so far as these are calculated to aid in extending our knowledge of the laws which respect the arrangement and distribution of animals. The systematist would thereby be enabled to judge better of the affinities of any new species, and to determine the place they occupy in Nature's plan, so intricate in its details, with more certainty and exactness. The collector, moreover, has it in his power to supply much information of the above kind, which, if not recorded at the moment, may be in vain sought for afterwards.

(9.) And let none think that, because we have so many works conducted on the plan of White's, and so much on record in these days respecting the habits of animals, there is nothing more to be learnt. Ray has remarked, that so rich is Nature, that a man born a thousand ages hence will still find enough left for him to do and notice.* The field open to the observer is really inexhaustible; and this is not more true in respect of the immense number of species inhabiting this globe, than of what is requisite to perfect the history even of those known. In how very few cases, if any, can we say that we have attained to a complete knowledge of any one species, so as to give a detailed account of all its characters and instincts, and the degree to which these are liable to be affected by an alteration in the circumstances of its life! To those who travel in foreign and remote countries, still more to those who are stationed in localities but seldom visited by man, the force of this remark must be obvious. But even without going from our own neighbourhood, or withdrawing from spots with which we have been long intimate, how much may be learnt in addition to what we yet know! It is not always the animals that we are most familiar with by name and frequency of occurrence, whose history we understand the best. And amongst the lower tribes especially, there are many that fall under our notice almost every day, respecting whose mode of production, food and instincts, we are as much

* *Philosophical Letters*, p. 35.

in the dark, as if they had been only just brought to our museums from a remote quarter of the globe.

(10.) Neither let us disdain to notice the most trivial facts that may be brought under our eye, from a feeling that they can be of no use. When White made the remark (at which some might smile) that the field-cricket " drops its dung on a little platform at the mouth of its hole," he probably did not stop to consider whether the fact were worth recording, but was actuated only by a desire to give the *whole* history of the little insect before him, so far as he could learn it. And it is impossible to say at the moment of what use the most trifling fact may be. It is impossible to determine the exact importance of any circumstance in the history of an animal until we *know* its whole history; and not only this, but the whole history of other allied species, so as to ascertain what is peculiar to the one in question, and what is common to all the species of the group to which it belongs. Many facts, which seem in themselves trifling, may be found hereafter of the greatest importance to science. They may lead to the unravelling of some knot, or the solving of some difficulty, that would long have remained a mystery without them. And one simple observation, thought nothing of at the time by the observer himself, may avail to the overthrow of an entire system, the fruit perhaps of years of labour and close meditation.*

* See *Cuvier, Hist. des Prog. des Sci. Nat.* tom. i. p. 5.

(11.) We have hitherto spoken of observations in Natural History, as made with a view to promote the advancement of that particular science, and to supply facts upon which all generalizations must rest. But some of these facts, when obtained, may be of use in other ways. They may assist in forming the statistics, as it were, of other sciences, particularly Meteorology, for which purpose they have been assiduously collected by some observers. The different times at which the periodic movements of animals take place, their times of breeding and hybernation, and many interesting phenomena of the same nature; also, in the vegetable kingdom, the times of the leafing of trees, and the flowering of plants, the ripening of fruits, &c. — all these are more or less connected with the progress of the seasons, and climatological considerations, and on this account are well worthy of our notice. The Calendar of Flora in the *Amœnitates Academicæ* of Linnæus is well known, and White's Naturalist's Calendar known, perhaps, still better in this country; and from the circumstance of the attention of the scientific public having been lately reawakened both here and abroad to this subject, we have been induced to devote to it a certain portion of this work, as a guide and stimulus to those who are disposed to join in making observations of this nature. What, however, we have chiefly to say on this head will serve as an immediate introduction to the Calendar, which we propose offering to the reader in another place. We shall simply state here that the regularity which pervades nature, as regards the recurrence of

periodic phenomena, is very striking. We mean
not that the incidents of the several kinds above
alluded to always fall out on a particular day,
though no doubt the averages of many years' observations, taken at intervals, would bring them to a
near coincidence. But it is rather the regularity
with which they uniformly succeed each other, from
which there is little deviation, that is so remarkable.
Why the toad should be always a few days later
in spawning than the frog ;—or why the pheasant
should hatch before the partridge, though the latter
pairs for the breeding-season long before we hear the
sexual crow of the former ; — or again, why the
apricot should invariably flower a few days before
the peach, yet generally come into leaf a day or two
later; these, and a thousand other little matters of
the same kind that might be mentioned, furnish
much room for reflection to the thoughtful inquirer.
Nothing also is more surely regulated by the seasons
than the various sounds emitted by different animals,
whether the notes of birds, or the cries of insects, &c.,
which fall so gladly upon the ears of the naturalist,
indicating to him the different feelings by which
such animals are actuated. The pleasure, indeed,
afforded by rural sounds has been often rapturously
spoken of by ardent lovers of Nature, and the poet
has left us lines on that subject, the beauty and force
of which have been frequently alluded to:—

Nor rural sights alone, but rural sounds,
Exhilarate the spirit, and restore
The tone of languid nature.
* * * * * *

> Ten thousand warblers cheer the day, and one
> The livelong night: nor these alone, whose notes
> Nice-finger'd art must emulate in vain,
> But cawing rooks, and kites that swim sublime
> In still repeated circles, screaming loud,
> The jay, the pie, and ev'n the boding owl,
> That hails the rising moon, have charms for me.
> Sounds inharmonious in themselves and harsh,
> Yet heard in scenes where peace for ever reigns,
> And only there, please highly for their sake.
>
> *The Task.*

But we are here speaking of the sounds of animals as simply indicating the progress of the seasons. It has been said, and it is not far from the truth, that if an observant naturalist, who had been long shut up in darkness and solitude, without any measure of time, were suddenly brought blindfolded into the open fields and woods, he might gather with considerable accuracy from the various notes and noises which struck his ears what the exact period of the year might be.* Some instances of a remarkable coincidence in this respect will be found mentioned in the body of this work.

(12.) There is much pleasure in watching and registering such natural phenomena as we last alluded to, whether, after all, we turn them to any account

---

* Humboldt has made a somewhat similar remark, with respect to the power of distinguishing the different hours of the day, in tropical countries, by the hum of the insects that succeed one another at fixed periods. He observes,

" The insects of the tropics everywhere follow a certain standard in the periods at which they alternately arrive and disappear. At fixed and invariable hours, in the same season, and the same

or not. Many persons also have found their chief happiness in a habit of observing the life and manners of the animals in their immediate neighbourhood, without any view to the facts so acquired being made subservient to the progress of Zoology. We would throw no hindrance or discouragement in the way of such observers. We desire not to say anything that might tend to check their inquiries, though no benefit were thereby to accrue to the higher departments of the science. For we are deeply sensible ourselves of the pleasure which attends an observing habit of mind, as well as its usefulness in other ways, besides its bearing upon the general objects of science. When a man has learnt to take an interest in the varied operations of Nature, which are everywhere being carried on about him, and has acquired the habit of directing his attention to such matters, and keeping his senses always alive to any new information thereby afforded him, he has made himself almost independent of outward circumstances. He has opened to himself a source of occupation and mental enjoyment, but little affected by the ordinary vicissitudes of life. Of how few of the pursuits of the world in general can this be advanced! How few can secure those

latitude, the air is peopled with new inhabitants, and in a zone where the barometer becomes a clock (by the extreme regularity of the horary variations of the atmospheric pressure), where everything proceeds with such admirable regularity, we might guess blindfold the hour of the day or night, by the hum of the insects, and by their stings, the pain of which differs according to the nature of the poison that each insect deposits in the wound." *Pers. Narr.* (Engl. Transl.) vol. v. p. 95.

who follow them from disappointment and ennui, or are of that nature that they can be carried on in every possible situation, without prejudice or inconvenience to others! The pursuit of Natural History is itself a relief from ennui, and from many of the unavoidable anxieties to which the human mind is exposed. There have been persons, who have been forced to keep residence in some of the most desolate spots on this globe, and who have declared that it was the study of Nature alone which made their condition tolerable. How well then must this study be calculated to augment our happiness in more favoured climes, like our own! We are not surprised, indeed, at some men not becoming professed observers; but sure we are that those whose inclinations lead them that way, have an enviable advantage over others. They have always a resource to turn to, for the purpose either of filling up those leisure hours, which hang so heavily upon many, or of diverting the mind from anxious and oppressive thoughts.

(13.) It is, indeed, to be regretted that there are not more observers than there are. Though all cannot become such, there are surely very many who might, and who in this way would add greatly to their usefulness as well as happiness. We speak not of professional men, whose time is necessarily engrossed by business, or avocations, which could not be neglected for scientific pursuits. But it is lamentable to think, as an excellent author has observed, how many " waste a whole life, without ever being once well awake in it, passing through the world like a heedless traveller, without making any reflections or observations, without any design or purpose beseeming a man."* There are some who are not tied down by their circumstances to any particular employment, and who are content to do nothing. They either let their faculties rust in idleness, or direct them to objects unworthy of rational beings. Might not, we repeat, such individuals fill up their time more profitably, and in a way more calculated to advance their own happiness, by applying their minds to the great Book of Nature, always open to them, but hitherto unread? Is it not to be regretted that they can find even no amusement in the exercise of their senses, in discerning and noting down what there is of marvellous and instructive in the works of the Creation? The very variety that exists in Nature, the endless diversity of plan, the marks of skill and adaptation that everywhere meet us, seem almost to put forth an irresistible claim to our regard, and to command our attention. And when we have it in our power, from our situation and circumstances, to direct our studies that way,—not to do so is, to say the least, to shut ourselves out from some of the purest pleasures of which the human mind is capable. But to pass from the case of these persons, we apprehend that there are some professed lovers of Natural History, who want to be further stimulated in the exercise of their own selected pursuits. Such as live at a distance from the great centres of literary concourse, who have no access to museums and

* *Lucas on Happiness.*

libraries, and who have few or none to consult, or
with whom they can associate, sometimes fall into
habits of indolence, from a feeling that they can do
but little for the advancement of science.* But we
have already said (9) how much is often within the
reach of these persons. And if they could only be
induced to become observers upon some regular
plan, they would soon find how little room for complaint there was on the above ground ; how, indeed,
objects, worthy of notice, crowded upon their view
in proportion as they sought after them, of which,
perhaps, some had been daily under their eyes, but
passed over, from the mind never having been properly directed to them.

(14.) The habit we are here recommending is peculiarly adapted to a country life; and, in the pleasures which it affords, offers a full compensation for
the loss of those advantages, which are only to be
reaped in the society of large towns. In the midst
of Nature's best gifts, and surrounded by the many
living inhabitants occupying the same ground as
himself, the observer hardly knows what it is to be
alone. He finds company wherever he goes, and of
a kind, moreover, from which there is always some
lesson to be gleaned, some instruction to be gathered. He feels the truth of the great poet's remark,
that—

" Solitude sometimes is best society."

Nor is there here any insipidity or sameness to tire
or disgust him. He takes the same rounds every

* See, on this subject, *For. Qu. Rev.* vol. xiv. p. 317.

year, and there is always something fresh to quicken his attention—some novelty to recompense his researches. There is, also, an inducement to take that regular exercise, which is so essential for maintaining the health of the body. We know how common it is for persons of sedentary habits to neglect this important matter, thereby inviting all those diseases which arise from torpor and inactivity. Or if they do force themselves to take daily what is called a constitutional walk, too frequently they derive little benefit from it, in consequence of the mind not participating with the body in the imposed task, but still recurring to the occupations at home. It is, then, a great advantage, in respect of health alone, not merely to have an inducement to go out, but, the moment we do so, to be able to let the mind expatiate freely on surrounding objects ; to have it in our power to convert every walk into a source of instruction as well as pleasure, and to find happiness where others seldom even think of looking for it.

(15.) As a further encouragement to the forming a habit of observing the works of Nature, we might mention, what has been so often alluded to, its tendency to foster, if not to generate, a devout turn of mind towards their adorable Author. Undoubtedly it has this effect, where there is no perverseness or viciousness of temper present to counteract it. In watching the habits of animals, and the provision made for their welfare and happiness, in noting their varied instincts, their arts and stratagems to obtain the necessary support for themselves and young, their mode of defending themselves against their enemies,

and all their ways so replete with matter for reflection and astonishment, we cannot but trace the
finger of their Great Creator: we cannot but consider all we see as affording the clearest indications
of His over-ruling Providence. It speaks to the
existence of some mighty Power, whose secret influence upholds order and harmony, amid what would
otherwise be a chaos of confusion and turmoil, from
the conflicting interests of so many different agents.
View the animal world in one light, and it would
seem to be a world of rapine and bloodshed. We
find different tribes waging a perpetual war against
each other,—the weaker daily falling a prey to the
stronger, until we might suppose that these last
would be left alone upon the earth. Yet it is remarkable, that, notwithstanding this constant struggle for the mastery, or what is more frequent, this
necessity of yielding to the attacks of a superior foe,
the relative numbers of each species of animal seem
on the whole to remain nearly the same. Certain
species may have disappeared from the face of the
earth, or from particular districts, through the agency
and interference of man, or from the operation of
other causes; but this is not ordinarily the effect
produced by one species preying upon another.—It
is further remarkable, that, though many animals are
thus in perpetual jeopardy of their lives from the
number of their enemies, they yet seem in the enjoyment of as much happiness, as if all were security
and peace. Who can witness the gambols of a hare,
proverbially one of the most defenceless and perse-
cuted of creatures, without being struck with this

reflection ?*   To speak of the frolics of a lamb, so familiar to every one, or of the evident signs of happiness in all young animals, would be quite superfluous.   But the same signs of happiness appear everywhere, even in the lower tribes of animated nature.   It will be remembered by those, who are conversant with Paley's Natural Theology, how he has adduced the instance of swarms of young shrimps, which he had observed jumping repeatedly out of the water, from no apparent motive but that of the feeling of happiness, as one of the many proofs of " the goodness of the Deity."   He considered it as speaking to the fact of His having not merely provided for the wants of His creatures, but

* The reader will probably remember Cowper's humorous lines in his " Epitaph on a Hare :"—

A Turkey carpet was his lawn,
  Whereon he lov'd to bound,
To skip and gambol like a fawn,
  And swing his rump around.

His frisking was at ev'ning hours,
  For then he lost his fear ;
But most before approaching show'rs,
  Or when a storm drew near.

Eight years and five round-rolling moons
  He thus saw steal away,
Dozing out all his idle noons,
  And ev'ry night at play.

I kept him for his humour's sake,
  For he would oft beguile
My heart of thoughts, that made it ache,
  And force me to a smile.

given them, in addition to what was necessary as a means of welfare, a capacity for enjoyment in the simple exercise of their limited powers and faculties.* And no right-minded observer can witness many similar instances of the apparent happiness of animals, without being led in like manner to meditate on Him who is the source of all good. For ourselves, we never witness, in particular, the broods of gnats that may so often be observed hovering in the air during still weather, successively rising and falling every instant, without thinking of Paley's shrimps, and being forcibly reminded of his conclusion, that " it is a happy world after all." Perhaps, however, it is in the instincts of animals that we see the strongest evidence of a Creator's care ; in the workings of a sagacity, apparently making near approaches, in some instances, to that of man, though too narrow in its operations, and too uniform in its results, to allow of our attributing it to the exercise of any reasoning faculties like his. On this field, which has been so largely treated of by many authors, we are not at present about to enter far. We will not stop to enumerate the numberless cases, in which animal instinct shews itself,—or to point out the exact line of demarcation, by which it

---

* We would strongly recommend the perusal of Paley's work to all young naturalists, especially the chapter on " the Goodness of the Deity," in which will be found some of the above arguments carried further into detail. The system of animals devouring one another, Paley well defends, by shewing it to be, under all circumstances of their condition, the most merciful provision that could have been made.

is separated from reason.* Sure we are, that to
whatever source we attribute the effects we witness,
their adaptation to the purposes, which the several
agents have in view, is very remarkable. And we
feel convinced that there is some other wisdom besides their own, to which we must turn for a true
explanation of the mystery, although we may not
immediately discern the Hand that guides and regu-
lates their deportment.

(16.) And surely he is but little worthy of the
name of a naturalist, who rests satisfied with this
conclusion, without making any deductions calculated to improve and benefit himself. If the divine
wisdom and goodness are thus manifest in the
works of Creation,† do we not see stamped on these
works the character and disposition of Him, who is
our own Maker? If, especially, so much vigilant
care is exercised on behalf of the beasts that perish,
is there not conclusive evidence in respect of the
same or greater care being felt for man, made after
the divine image? There is no immediate impress
made upon us, as on the lower animals, to keep us
in the right way, because of our superior faculties

* Authors generally, who have written on this subject, seem
not sufficiently to have distinguished between intelligence and
reason. The former is found in many animals, and, when present, is generally in an *inverse ratio* to instinct; but reason, properly so called, is confined to man. See, in further illustration of
this point, Coleridge's " *Friend*," vol. i. Essay, 5.

† The Apostle tells us that,—" The invisible things of God
from the Creation of the world are clearly seen, being understood
by the things that are made, even his eternal power and God-
head." *Rom.* i. 20.

C

which render such a step unnecessary: nor is there any control exercised over our minds, to protect us from harm, that would interfere with our moral freedom, of which animals are altogether destitute. But then these higher gifts, which thus far more than supply the place of instinct, are themselves the proofs of what the Creator has done for man. They testify, in a still more remarkable manner, what is purposed by the Almighty in respect of ourselves, and speak to the more exalted kind of happiness to which he would lead us. What then must be the state of that heart which reaps no moral lesson from the great truths which Nature everywhere proclaims ; —which can observe the instincts and habits of animals, without thinking of the great unseen Agent, whose will they obey, or without seeking to inquire what that will is as regards himself, and which he is left (partly by the aid of his reason), to carry out or neglect, to his own eternal happiness or misery? As we said before, there must be some wilful perverseness or viciousness of temper present to counteract this effect. But where this is the case, it must be met by further and different arguments from what would be in keeping with this work. Instead, therefore, of following up this subject, let us express a hope that the supposed case is not of frequent occurrence, and that natural observers are generally awake to a sense of the moral and religious inferences, which ought to accompany them in their pursuits. We well remember once seeing a naturalist involuntarily shed tears, on the occasion of a hen partridge practising its well-known stratagem to divert attention from its young : she came running out of a ditch towards him, and threw herself at his feet, at the same time feigning herself hurt, and uttering a shrieking note as of the most poignant distress. He was touchingly affected by this artful device on the part of the poor bird, serving to show her maternal affection, and anxiety for the safety of her brood. Yet we know this device is always resorted to by the partridge in similar cases ; and the feeling, consequently, thereby generated in the breast of the pious observer, is not so much directed to the bird itself, as to that Providence, which by an instinctive impress teaches the lesson, and guides her as to the exact moment in which it is to be repeated.

(17.) Before we quit this part of our subject, it may not be improper to make one remark as a caution necessary to be impressed on some observers. Good and excellent as are the feelings towards the Creator, which the study of Nature is calculated to produce within us, we must not attach so much importance to them, as may lead to the exclusion of other considerations of far higher and weightier moment. In other words, we must not mistake natural religion for revealed ; nor suppose that the truths, which this last alone can teach us, are to be learnt by the most attentive regard to this lower world and its varied productions. This is a solemn reflection, which it would be out of place to follow far here. And we simply allude to it, from our belief that there are some naturalists, who make their study of the works of the great Creator the whole of their religion, and the sense they thereby attain of His adorable perfections a plea for neglecting the teaching of His revealed Word.—May we never forget, that both these come from the same author ;—that man's welfare and interests are as much concerned and bound up in the one as in the other ;—and that no feelings are to be trusted, which prompt us to make our own selection, in respect of what is to be our rule of duty or our guide to happiness.

(18.) In addition to what has been advanced in favour of habits of observing, as connected with the study of Natural History, there are some collateral advantages resulting from them, which deserve to be mentioned. Such habits tend generally to sharpen the senses, and to make them more ready at command, when wanted, whatever be the particular object soliciting our regard. Even he, who confines his attention to the science we are considering, and who never quits the field, wide indeed in itself, which nature opens to him, will after a time be surprised at the quickness with which he notices things compared with what he did formerly. The more he observes, the more he finds to observe. From having been long accustomed to have his eyes and ears always open, he perceives objects, and catches at sounds, which formerly he would have entirely passed over, and which others, less practised than himself, with difficulty distinguish, even when their attention is expressly drawn to them. The readiness with which the experienced ornithologist will hear and distinguish the notes of different birds, all singing together on a fine spring morning, is an instance in point; and we mention it from having sometimes found it almost

impossible to get an accompanying friend, on such
an occasion, to be sensible of some note, uttered
by a particular bird, perhaps close at hand, which
was to ourselves perfectly audible. The attention,
also, of an habitual observer is more and more arrested by little things, which formerly, though they
might have been noticed, would have made no impression, nor called up a single reflection in the
mind. We remember once being much struck with
the acuteness in this respect of a great philosopher,
whose senses, perhaps, were more practised and
awake than those of almost any person that ever
lived, and who was no less close and accurate an
observer in natural history, than in those higher
departments of science, to which he more particularly devoted himself, and which he cultivated
with so much success.* Walking with him one
day in the country, and discoursing on the different objects that fell under our view, his eyes
turned suddenly to the ground, his attention being
arrested by the appearance of some minute specks,
scattered here and there, which he was not satisfied
till he had taken up and examined. They proved
to be only the excrement of some small caterpillars feeding upon a tree overhead, from the branches

* The allusion made above, is to the late Dr. Wollaston; of
whom it has been well observed in a recent publication, "*Inerat
etiam Wollastono ea perspicacitas, ut quæ communi hominum sensui
parum obvia essent, ea statim animo arriperet atque complecteretur.*"
(Dr. Daubeny's Harveian Oration, 1845, p. 11.) This remark
referring to his quickness of mind, might have been quite as justly
applied to the quickness of his senses.

of which the atoms in question had fallen. But the manner in which, for the moment, he threw his whole mind into what would have been to others so insignificant and almost invisible an object, was strikingly characteristic of such an observer: the remarks, also, to which it led on his part, showed how the greatest intellects will consider nothing unworthy of their notice, from which the smallest addition of knowledge, loved for its own sake, can possibly be derived.*

(19.) But the naturalist, whose senses have been sharpened in this manner by habits of observing, will thereby have the advantage over others in matters not immediately connected with his own pursuits. He becomes a more quick and ready observer on all occasions. An eye that has been well disciplined, and accustomed to exercise, and which is always on the look out for information, is prepared to make the most of every opportunity of getting it. Two persons shall enter an exhibition together, and in the same time one shall have seen far more of it than the other, shall, perhaps, have traversed the whole in a general way while the other was only gazing vacantly about him, besides, here and there, noticing details which escaped the other's observation altogether, though occurring in that particular quarter to which he professed to have directed most attention. This is solely the effect of a habit of observing; a habit of knowing how to use the eyes

* The nature of the remarks above alluded to, in reference to the excrement of caterpillars, will be found further treated of in the body of this work.

when wanted, of adapting them to receive an immediate impression from the objects presented to them, of distinguishing minute differences, where others see none at all, and, when restricted in time, gleaning all that a momentary glance will allow. The case just supposed is in fact but another illustration of the lesson taught us in Mrs. Barbauld's well-known story of " Eyes and no Eyes " in her *Evenings at Home*. We all remember the moral of that lesson; yet how few, nevertheless, carry through life the spirit of the young active inquirer there represented to us, who saw so much where his brother had seen nothing, and whose interest was excited, and his senses kept on the alert, by what to the other afforded neither instruction nor entertainment. How many are there, rather, who the further they advance in the world, have their senses more and more blunted by disuse, and who not only regard half what they see (if they can be said to see at all) with apathy and indifference,—but even in cases in which they are more immediately interested, scarcely know how to employ their eyes aright, so as to profit by them as much as possible, or to ensure their being trust-worthy interpreters of whatever is submitted to their notice.*

RULES FOR OBSERVING.

(20.) The remark with which we concluded the last paragraph leads to the consideration of a few

* " It is surprising how little we see until we are taught to observe." *Forbes in Edinb. New Phil. Journ.* vol. xxxii. p. 85.

points, which it may be useful to impress on those
who purpose to become observers of nature. And
first as regards the importance of *observing accu-*
rately. The value of the facts which we propose to
collect must rest primarily on the correctness with
which they are noted. And though it may seem
hardly necessary to insist on this, yet in fact it is so,
from the numberless errors into which persons are
led every day, in respect of the objects that fall
under their notice, from not giving them a sufficient
examination. If we are not habituated to looking
closely at objects, we are apt to be misled by first
appearances, and to mistake one thing for another.
Thus we have known many persons mistake the
" large hop-like fruit" of the wich-elm, so conspicuous in the latter part of the spring, for leaves;
and they would scarcely believe that the tree was not
in full leaf, until it had been made evident to their
senses, by plucking a branch, that what they saw was
merely the expanded seed-vessel, the true leaf-buds
being as yet scarcely evolved. Had one of these
individuals been requested by a botanical friend to
register the dates of the leafing of the different trees
in his neighbourhood, he would have been liable in
this instance to have registered a false fact, without
the slightest suspicion of its being incorrect.   In
like manner the fall of the floral leaves, or *bracteæ*,
of the lime, towards the end of August or beginning
of September, might lead a hasty observer to imagine that the true leaves were beginning to fall at
that early period. How often again are the great
flies, which enter our apartments so frequently in

the autumnal season, and keep up a perpetual humming not unlike that of a drone, consider-ed to be bees.* Indeed it is hardly possible to enter much into society, without being struck with the positive manner, in which we sometimes hear persons stating matters as facts, which every one, in the least degree acquainted with science, knows must be erroneous. We remember once hearing a gentleman distinctly assert that humming-birds were found in this country, and surprised at our denial of the fact: it appeared that he had frequently seen the well-known humming-bird hawkmoth† hovering over the flowers of plants, and had mistaken it for a real bird. Such errors are pardonable in those who have paid no great attention to natural objects, and who pretend not to collect facts for scientific purposes; but they show the caution that must be exercised, and the scrutiny that must be made, before trusting to first impressions upon the senses, as soon as we may have enlisted ourselves in the rank of scientific observers.‡

(21.) But besides being carried away by first im-pressions, we must carefully guard against *being in-fluenced by prejudices*. And such prejudices as are likely to interfere with correct observation are of two kinds. We may be prejudiced by popular no-tions, which have been long and generally enter-

* Allusion is here made to the *Eristalis tenax* of Meigen.

† *Macroglossa stellatarum*, Steph.

‡ We may allude to the erroneous statement of Jesse, that "ear-wigs turn to flies," (*Gleanings*, 3rd. Ser. p. 149,) as an instance of the mistakes which sometimes occur in our popular works from incorrect observation.

tained, and which are originally due either to igno-
rance or superstition ; or by a bias affecting our own
minds individually, and caused by a peculiar train of
thinking into which we have fallen. The first class
of prejudices are those which prevail so frequently
with persons of slender education and weak abili-
ties, who readily embrace whatever notions have
been handed down to them by their forefathers,
without troubling themselves to inquire whether
they are correct or not. These hindrances to the
diffusion of true science are gradually dying away,
as knowledge spreads, and education advances,
and we hope in a few years most of them will be
extinct. But they still in many instances maintain an obstinate hold over the mind, utterly in-
capacitating it for investigating the truth. When
due to superstition, they are, as White observes,
sucked in, as it were, with our mother's milk, and
growing up with us at a time when impressions
generally take the most lasting hold, become so
interwoven into our very constitutions, that the
strongest good sense is required in order to dis-
engage ourselves from them.* Every one's memory
will supply them with instances of this class of prejudices, of such frequent occurrence in respect of
matters of Natural History, that it is needless to
particularize. But we guard all persons, wishing to become observers, against their influence.
We require them to dismiss every notion from their
minds, imbibed from others, until they have taken
correct steps to verify it, and to go forth into the

* Letter xxviii. to Daines Barrington.

haunts of nature, determined to see and judge for themselves, with as much caution and exactness as possible.

(22.) The second class of prejudices above alluded to are not those of an uneducated mind, but of one that has been *warped by long reflection on a peculiar train of ideas.* Under such circumstances, the observer sees everything through a distorted mirror. His usual habits of thinking interfere, and mix themselves up, with the impressions of the senses, substituting, in many cases, a false image for the true one. Nothing is noticed but what bears upon the favourite matter of speculation; or, if observed, it is not observed in its proper colours, and often, when the mind is on the search for novelty, it is so full of a preconceived idea, that it fancies it sees what has no existence in nature. This is too frequently the case with those naturalists who have some theory to maintain, and who seek to uphold and confirm their particular views by an appeal to facts. Everything appears to fall in exactly with their preconceived notions. Instances in point are readily found; sometimes resemblances traced, and inferences drawn, the former quite inappreciable to other eyes, the latter quite illogical to other minds. There is not here, any more than in the case of the class of prejudices first mentioned, any real intention of perverting the truth; but the judgment is so blinded by a strong desire to find things squaring with its own views, that it becomes utterly unable to take proper hold of the naked truth, or interpret it aright.

(23.) To those who are collecting facts for their own use, and not merely to supply others with materials for generalization, it may be useful to be taught the importance of observing a particular fact *in a sufficient number of instances,* before coming to any conclusion respecting it. Long and repeated observation is required in most cases to enable the mind to generalize correctly. And this, which is true in all sciences, is especially so in the science of Natural History, in which there are as yet so few principles well founded and universally received. In proportion to the advance of our knowledge of such principles, will be our safety in taking analogy for our guide. Yet analogy is rarely to be trusted, as those will testify, who have studied longest the book of nature, and searched it the most deeply. We often find quite a different law (if we may use the term) prevailing in two cases which at first view appeared parallel. By law we mean some rule which seems to regulate either the structure or the habits of animals. Let us take the instance of aquatic and land animals. What observer who was acquainted with the mode of respiration in the land mammals, and then found fish breathing by gills, would at first suspect that the cetaceans accorded with the former in the organs adapted to the exercise of this function. And in the case of certain groups lower down in the system we may observe the converse of this singular phenomenon. The land crabs breathe as exclusively by gills, as any of their congeners which reside always in the water. So also, it is now found, do the common wood-lice, though for a long time this was never suspected.* Again, when first it was observed that the natatorial birds had webbed feet, doubtless there was a feeling that this structure and the power of swimming went always together; yet every tyro now knows that there are birds with webbed feet which never swim, as there are others, in which the toes are divided to their origin, which swim admirably. In fact, numberless are the cases in which we find a deviation from some general rule. Nature seems to delight in the production of apparent anomalies, for the express purpose of baffling our attempts to bind her by the fetters of a preconceived system. These are for a while so many stumbling-blocks in our way, which we as little expect to meet with beforehand, as we are prepared to explain them when they occur. There are some facts, indeed, of this kind so extraordinary, that had they been recorded formerly, when correctness in observing was less attended to than at present, and had they rested only on such statements, they would not now be received. And some which were so recorded, and for many years disbelieved on account of their strangeness, have been observed again in modern times, and to the surprise of naturalists, been proved true.† From these circumstances we learn caution, as well in pronouncing any fact in itself to be true or false, as in

* See Duvernoy and Lerebouillet on the organs of respiration in the *Crustacea Isopoda* in the *Ann. des. Sci. Nat.* Ser. iii. tom. xv. p. 177.

† As an instance in point, we may mention the fact of some of the great serpents in India incubating their eggs. See a note on this subject in our edition of *White's Selborne*, Lett. xvii. to Pennant.

drawing any inference from facts, which, though true, have been noticed but in a limited number of instances. We must enter upon the study of Nature prepared to believe everything which is not actually impossible, and which has not certainly been disproved; ready to confirm the observations, however strange, of those who have gone before us, if occasion offer; ready, also, to give up what we had considered our best established principles, if called upon to do so by a more extended research. In a word, we must equally repress a harsh judgment upon the statements of others, and a too hasty desire to generalize on our parts.

(24.) Besides observing the same fact in a sufficient number of instances, we should, as far as possible, observe it in instances, in which *the attending circumstances are different.* This is especially of consequence in all that relates to the habits of animals, and the dates of periodic phenomena; for these are often so modified by some peculiarities of place or season, as to offer quite a different result from what would ordinarily occur. To mention many examples of either of these two classes of exceptions would be unnecessary. As regards the manner in which instinct often conforms to circumstances, we may just allude to the case of birds choosing such materials for their nest as are most readily obtained, or adapting the situation of it to the nature of the particular locality.* The altered

* See *White's Selborne,* Lett. 56, to Daines Barrington ; also Lett. 21 to Pennant, on the subject of daws breeding in unlikely spots. See also Bolton's *Harmonia Ruralis,* vol i. p. 492.

habits, in many instances, of animals in confinement, or in a domestic state, are notorious. As regards dates, of which the Naturalist's Calendar is almost exclusively made up, they are liable to be influenced not merely by season, but by many accidents affecting the individuals to which they relate. Age, health, sex, the interference of man, an unusual supply of food of a particular kind in certain spots, the greater or less prevalence of enemies, and several other causes, may all operate in forwarding or postponing any natural occurrence. Hence, if we desire to arrive at the true law regulating any class of observed facts, especially when some of them appear to deviate from the usual order of things, we should carefully note down all the attending circumstances; all such at least as can by any possibility have any influence over them. And herein it is much better to be too diffuse than too meagre. We must guard against thinking, in any doubtful point, that it can be of no use to note this or that circumstance, for we are not at the moment in a position to determine its importance. This can only be decided afterwards, when we or others shall have collected many such parallel cases; and we can see what circumstances are common to them all, and what are evidently adventitious in a few only. And if we neglect to register at the time all that the eye notices, we may in the end deeply regret the omission of many particulars, when it is out of our power to supply the defect by any fresh observation. For unlike the natural philosopher, who can often, by experiment, create for himself all the circumstances attending a particular

phenomenon, or vary them as he may choose, and so
repeat his observations as often as he will; — the
naturalist has to collect facts, which are generally
beyond his control, and for the occurrence of which,
he must wait patiently till Nature herself bring them
under his notice. It is hardly necessary to add, how
important it is that we be accurate as well as cir-
cumstantial, in our detail of the particulars accompanying any fact; and that our record of them be
registered at the moment, without trusting to
memory, that fruitful source of error and exagge-
ration.

(25.) It may be useful to young observers, to say
a few words in reference to the best manner of
searching for facts in Natural History.—When a
traveller reaches a new country, or one which has
been seldom visited by naturalists, almost everything
he meets with will be new also, and at every step
he is called upon to register some fact or other.
There are here scarcely any directions necessary
beyond what have already been given. He must
have his eyes and ears open, and with pencil in hand
faithfully record whatever falls under his observation. Often too in unexplored regions, there is no
difficulty in watching the habits of animals, from
their being unaccustomed to man, and under no fear
at his approach. Where this is the case he possesses
great advantages, and has an opportunity of seeing
more of their ways, as well as seeing more surely.
But this seldom happens anywhere; and in thickly
inhabited countries the observer must exercise great
caution and address, or even have recourse to strata-

gem, in order to become acquainted with the objects of his search. Of course this remark is not applicable to the collecting of facts respecting the lower animals, which are often stationary, and quite indifferent to being watched. Here the observer may immediately resort to those localities, which are tenanted by the particular tribes he wishes to study. But as regards quadrupeds and birds, with which principally the greater number of persons occupy themselves, they are generally not to be approached without stratagem: and if they are species frequenting plains or other open places this is unavoidable. Also, if the observer is collecting facts for the history of any particular species, he must follow up that species wherever it is to be met with, and must exercise care and ingenuity, as well as a perpetual watchfulness, in getting acquainted with all the particulars of its life. But if our researches are carried on in a woodland district, especially in our own country, and we are not particularly interested in one species of animal more than another, we would strongly recommend, or at least as an occasional practise, the taking our station in some particular spot, and, instead of going to look for objects, noting down whatever occurs, which we consider worthy of our regard. Sometimes in this manner we shall have our attention called to little matters, which, by being always on the move and on the look out for more striking facts, we should have long continued to overlook. In the dearth of larger animals, or in the intervals between their drawing near enough to be observed, we shall naturally look closer into what is

at hand, if it is only to while away the time. And
there is no spot so barren of life as not to afford
insects and other minute creatures, whose ways and
actions will amply reward the curiosity of such as
deign to bestow their attention upon them. Many a
naturalist, when standing under a tree to shelter
himself from a passing shower, has been led to the
observation of facts, connected with the smaller
animals, or the more concealed processes of vegetation, or perhaps to the discovery of some new species
of lichen or fungus, which, but for this accident, he
might never have noticed. And to return to the
case of the observer in the wood, he will gain an advantage over the larger animals by being himself
stationary. It often makes a considerable difference
whether you approach them, or they you. They are
much more easily frightened in the former case than
in the latter; and less easily induced to resume their
freedom of manner, after the interruption caused by
having their retreat suddenly broken in upon. Add
to which, there is always the chance, when moving
about, that we disturb some bird or quadruped,
which, till the moment of surprising it, we did not
observe, and which not only takes the alarm itself,
and instantly hurries away to be seen no more, but
at the same time gives the alarm to others, and
causes a general desertion of the locality we are
traversing.

(26.) Perhaps it will be thought that to act upon
the plan just suggested, requires *much patience on
the part of the observer:* but we fear without patience
a man will never be a proficient in Natural History,

more than in any other department of science; at least he can learn but little of the habits of animals from his own autopsia. And perhaps it is not sufficiently known or considered how near it may be possible to get even to the most timid animals, to watch them in their actions, if the observer will be occasionally content to remain still and motionless for a few minutes. We have seated ourselves in a wood, and, while keeping perfectly quiet, without moving a limb, have had the hares sporting at our very feet, as if quite unconscious of our proximity: the same thing has occurred with the water-rat, one of the shyest of our native quadrupeds, and which in general darts into the water with great rapidity on the slightest alarm. It is moving objects or the noise of some one approaching, which most readily frightens animals. Yet even where it becomes necessary to advance, in order to see anything of their ways, as where they are feeding at a distance in open ground,—we may sometimes, by dint of great caution and patience, get almost completely up to them, without causing them to fly. We must only be careful to take very short steps, and at intervals, always desisting the moment our object shows any apprehensions, and remaining stock-still, till we see it resuming its former state of ease, and returning to its food, or to whatever else it is occupied with. By these means, we remember once succeeding in actually getting so close to an old rabbit feeding upon a lawn, as to secure it with a common walking-stick, where there was nothing whatever to conceal our approach to the animal, which appeared in perfect

health and nowise disabled. It might not be easy to
do this again; and some accidental circumstance,
besides patience on our part, might have favoured
the success which attended the experiment in this
instance ; but certainly *without* patience, it would
have been altogether impracticable ; and we mention
it in order to show, how it is possible gradually to
habituate animals to the appearance of man, just as
we know birds get in time habituated, as the farmer
is too well aware, to the scare-crow in the fields.

(27.) There is but one hint further, which we
would suggest in addition to those which have been
already given; and that is, a recommendation to
observers to *visit the same spots repeatedly*, especially at different periods of the year. To those
whose situation and circumstances necessarily limit
their researches to a small field, it will be an encouragement to be told of the fruits that may be
gathered from attending to this plan. Having once
gone over the ground, or visited it perhaps a second
season, and gleaned nothing from it which they had
not observed before, they might be inclined to think
their task done, or rather we should say, their
amusement ended. But we assure them this is not
the case. We advise them to return again and
again to the same localities, and this at all seasons;
and we feel assured they will find in the end their
trouble repaid by the discovery of some new fact,
their curiosity further stimulated. To those who
are collecting materials for a Fauna or Flora of
any district, it is notorious, not only what a constant succession of different species are to be found

in a given neighbourhood as the year advances,* but how occasionally in certain seasons some species will abound, and then disappear, perhaps, and not be seen again for several years. We remember once finding several species of orchis growing abundantly in a pasture, which we had been in the habit of crossing again and again in former seasons, without noticing one; and two of these have not been observed since in that locality. But every botanist, and yet more every entomologist, can adduce similar cases. Even among the larger animals, novelties will perpetually occur; and we cannot do better than call to the reader's mind, in reference to this subject, the remark with which White begins his letter to Daines Barrington, giving an account of that rare bird the stilt-plover, which he had just procured in his own neighbourhood. He says, " It is now more than forty years

* St. Pierre affirms, — " I can say, with truth, that I have not permitted a single day to pass, without picking up some agreeable, or useful observation."

Afterwards, in illustration of the boundless extent of Nature, he writes thus :—" One day, in summer, while I was busied in the arrangement of some observations which I had made, I perceived, on a strawberry plant, which had been accidentally placed in my window, some small winged insects, so very beautiful, that I took a fancy to describe them. Next day, a different sort appeared, which I proceeded likewise to describe. In the course of three weeks, no less than thirty-seven species, totally distinct, had visited my strawberry plant : at length, they came in such crowds and presented such variety, that I was constrained to relinquish this study, though highly amusing, for want of leisure." *Studies of Nature*, (Engl. Transl. by Hunter,) vol. i. p. 2.

that I have paid some attention to the ornithology of this district, without being able to exhaust the subject: new occurrences still arise as long as any inquiries are kept alive." * In fact, as we observed in a former part of this introduction (9), the subject is inexhaustible; and it is a satisfying thought to know that we may continue all our life long in a retired village, like the above amiable observer, to whom we have so often alluded, without our entertainment coming to an end; to be assured that Nature will spread a fresh repast for us each year, and that so long as we keep up a relish for her works, she will take care that our interest therein shall not flag for want of variety. It is no little solace to have an occupation which will hold by us, as White tells us it did by him,† even to declining years; —which may tend to quiet and compose our minds under the infirmities of age; and even serve a still higher purpose, when accompanied by those religious feelings which it ought continually to strengthen, and when aided by the help which Revelation so bountifully affords; thus preparing us for our last great change, and raising our thoughts from this lower world to Him who made it, " In whose presence is fulness of joy, and at whose right hand are pleasures for evermore."

* See Lett. xlix, to Daines Barrington.
† See the concluding portion of the Preface to the *Natural History of Selborne.*

OBSERVATIONS

IN

NATURAL HISTORY.

The works of the Lord are great, sought out of all them that
have pleasure therein.

Ps. cxi, 2.

———— Birds and Beasts,
And the mute Fish that glances in the stream,
And harmless Reptile coiling in the sun,
And gorgeous Insect hovering in the air,
The Fowl domestic, and the household Dog,
In his capacious mind —— he loved them all.

WORDSWORTH.

OBSERVATIONS ON QUADRUPEDS.

_____

MISCELLANEOUS.

THE most common occurrences, and such as are brought under our eyes every day, sometimes escape the notice of inobservant persons. A farmer, who had lived all his life among stock, was not aware, till I drew his attention to the fact, that horses and oxen rise from the ground differently. There is a slight difference in their mode of lying down, the horse not generally remaining so long upon his knees as the ox, before bringing the rest of his frame to the ground. But in getting up, the horse invariably rises first upon his fore-legs, before rising upon his hind. The ox, on the contrary, rises first upon the hind, and often remains upon his knees some few seconds until his hind legs are straightened These differences probably prevail throughout the two Cuvierian groups of *Pachydermata* and *Ruminantia*, to which the horse and ox respectively belong. The elephant and rhinoceros both rise first upon their fore-legs, like the horse; so does the pig : the sheep, goat, and deer, in this respect, are like the ox.

The horse, in trotting or walking, lifts his feet off

D

the ground in a certain order : first he raises the off
fore, then the near hind, then the near fore, and lastly
the off hind.   The appearance, as is well known, is
that of the two legs which are diagonally opposite
being raised nearly simultaneously; but the two on
the same side following one another at a moderate
interval, the hind one advancing first. The elephant,
as many observers have noticed, appears, in walking,
to move the two legs on the same side at the same
time ; and it has occasionally been thought that the
order in which the legs are raised from the ground is
different from that in the horse. But, upon close
watching, it will be seen that this order is in all
cases the same; the only difference consisting in
the length of the intervals between taking the feet
successively up. In the elephant, the interval between raising each hind-foot and the fore immediately in advance of it is very short, and it becomes relatively shorter as the pace increases.
When the animal walks very slowly, the legs appear to move just as in the horse ; the interval in
the two cases being the same. The same may be
observed in the rhinoceros, though I have had no
opportunity of noticing this animal moving fast, so
as to say whether it then resembles the elephant
in the appearance of the legs or not. The giraffe,
whether it walks fast or leisurely, appears to move
the two legs on the same side together, as in the
elephant. It is observable, that both the giraffe
and the elephant have short bodies (the former
especially) in respect of their height and length of leg.
Whether this has anything to do in lessening the

interval above spoken of, I am not prepared to say; but I thought I observed, when the rhinoceros was walking slowly, in which animal the body is rather long in comparison of the legs, that this interval was also *longer* than in the horse, as it is *shorter* in the two animals just mentioned.

———

Dogs, cats, and pigs lie down, in general, to suckle their young. In one instance, however, I saw a sow suckling a whole litter of pigs at once in the standing position. The sheep and cow, and probably all ruminants, as well as the mare, stand for that purpose. Mr. Yarrell informs me, that the mode of suckling in the giraffe, as observed by him in the case of the two young born in the Zoological Gardens, London, is the same also, and precisely similar to that of the mare.

———

White has observed that " the brute creation recognize each other more from the *smell* than the *sight*; and, in matters of identity and diversity, appeal much more to their noses than to their eyes." In confirmation of this, he mentions the well-known circumstance of ewes and lambs not being able readily to distinguish one another after they have been shorn; and the confusion that there is, in like manner, in the flock after sheep have been washed.*

* See his *Naturalist's Calendar, with Observations in various branches of Natural History;* by Aikin, p. 92.

Many other facts of a similar kind might be adduced, tending to show the correctness of this remark. Thus, to take another instance from sheep, it is a common practice with shepherds in the lambing season, when they have occasion to put a strange lamb to a ewe that has lost its own offspring, to clothe the supposititious young one in the skin of the dead lamb of the mother to whose care it is to be confided. Without this precaution, the ewe would refuse to suckle it; but the former does not seem aware of the fraud thus put upon her, so long as the *notus odor* (as White terms it) is present to direct her judgment in distinguishing, as she conceives, her own offspring from that of another.

It is also very remarkable to what a distance animals will discriminate, by the smell alone, others of their own species; sometimes long before they are in sight of one another, and when therefore the recognition is manifestly due to the above faculty. The unaided sight, on the contrary, seems in many cases to make but little impression upon them. This is attested by the circumstance of their not recognizing in general the most faithful representations of their own species on paper or canvass. I had a cat which passed by with the most sullen indifference a worked portrait of a cat sitting upon a stool, when purposely placed in her way, though so well executed as to deceive one or two persons on their entering the room for the first time where the portrait was. Yet it is to be remarked, that animals do not in all cases remain unmoved by pictures of other animals *not* of their own species; for a relative informs me of an instance in which she was present when a dog that was much afraid of cats shrunk back, and showed manifest signs of fear, on seeing a picture of a cat, that happened to be upon a low screen in the room of a gentleman upon whom his master was calling, and whom he had accompanied into the apartment. Animals will also sometimes be deceived by representations of what constitutes their usual food, provided it be not prey which is hunted down by the scent.*

* Humboldt mentions an instance of this in his Travels in America. Speaking of the *titi* of the Oroonoko (a species of monkey) he says, " It is extremely fond of insects, particularly of spiders. The sagacity of this little animal is so great, that one of those we brought in our boat to Angostura distinguished perfectly the different plates annexed to the *Tableau Elémentaire d'Histoire Naturelle* of Cuvier. The engravings of this work are not coloured ; yet the *titi* advanced rapidly its little hand in the hope of catching a grasshopper or a wasp, every time that we showed it the eleventh plate, on which these insects are represented. It remained in the greatest indifference when it was shown engravings of skeletons or heads of mammiferous animals."

But, in a note upon the above passage, he makes a statement confirmatory of what has been said above, as to the general indifference of animals to pictures of other animals in cases where they would be ordinarily guided by the nose rather than by the eye. He says, "I shall observe on this occasion, that I have never heard of a picture, on which hares or deer were represented of their natural size, and with the greatest perfection, having made the least impression even on hunting-dogs, the intelligence of which appeared the most improved. Is there an example (he adds) well ascertained of a dog having recognized a full-length picture of its master ?—In all these cases, the sight is not assisted by the smell."—*Personal Narrative*, (Eng. Transl.) vol. iv. p. 527-8.

Happening, on one occasion, to throw a dead long-eared bat to a cat, the latter was observed to seize it with the greatest avidity, at the same time uttering a kind of savage growl (as cats often do when they capture a favourite prey); and presently retiring with it to a corner, soon devoured it entire, not rejecting even the flying membranes. This was a well-fed, parlour cat, and apparently not suffering particularly from hunger. On repeating the experiment with another cat, this last took not the least notice of the bat whatever, though repeatedly placed in its way. This shows how different individuals of the same species will occasionally differ in respect of food; and how little importance is to be attached to the fact of any particular food being selected by an animal in a single instance. Thus I have seen, in some botanical books, mention made, in the case of certain plants, that " a horse ate it," or " a cow refused it;" when perhaps, on a second trial, another horse would have refused it, and another cow have eaten it.

It is probable, also, that all animals, from hunger, or some other cause, will deviate occasionally from their usual food. We have an example of this in the case of the hedge-hog that devoured a snake, recorded by Mr. Broderip;* another, in that of the polecat which has been known to prey on eels.†
Owls, rooks, jackdaws, and carrion-crows have all

* Zool. Journ. vol. ii. p. 19.
† Bewick's Quad. (edit. 8.) p. 253. The stoat also sometimes devours eels : see a notice of this fact further on in this work, communicated to me by Mr. Selby.

been observed to devour fish, in one or more instances: * on the other hand, the heron, which usually preys on fish, is well satisfied at times with the common water-rat, which, according to Mr. Waterton, it gulps down entire. † Many more such anomalies are probably on record, or might be observed if sought for, in the varying habits of different animals. I will only add here, that though, as is well known, owls and cats usually reject shrews, I have known both occasionally to devour them, especially kittens, which have not as yet had much opportunity of satisfying their appetite for mice.

BATS.

THE different species of bats found in this country appear, for the most part, to have each of them their own peculiar place of resort. With us the noctule ‡ and Natterer's bat § occur chiefly in trees, but the former sometimes also in houses. All the specimens found by me of the whiskered bat, ‖ and they have been several, were taken in rooms in

* See *Atkinson's Compendium of the Ornithology of Great Britain*, p. 23; also several notices in *Loudon's Magazine of Nat. Hist.*, all speaking to the fact of owls eating fish. Instances of the rook and jackdaw feeding on fish will be found mentioned in a subsequent part of this work. For a case of the carrion-crow's doing so, see *Ann. and Mag. of Nat. Hist.* vol. xv. p. 168.

† *Loud. Mag. of Nat. Hist.* vol. viii. p. 455.

‡ *Vespertilio noctula*, Gmel.

§ *V. nattereri*, Kuhl.

‖ *V. mystacinus*, Leisl.

houses. The long-eared bat* resorts in great numbers to churches, and the roofs of houses; in which
last situation they may often be observed in great
numbers, congregating in the angles formed by the
meeting of the timbers, and hanging in clusters
of twenty or thirty together, like swarms of bees:
I have once or twice only taken solitary individuals of this species from hollow trees. The pipistrelle † delights especially in the crevices of decayed
brickwork, in the cracks of old gateways and door-
frames, or behind the leaden pipes frequently attached to buildings for carrying off the rain. In
these situations they sometimes collect in astonishing quantities, several hundreds having been counted
coming out, one after another in quick succession,
from the same hole on a summer's evening.‡

Noctule.—On the 14th of June, 1842, I received
from Mr. Henderson, of Milton Park, Northampton-
shire, three old female noctule bats, one having two
young ones at the breast, and each of the others one.
They were taken, he informed me, from a hollow
tree. The young hung at the teats, the mother
wrapping them round with her wing, so as almost to
conceal them from view: occasionally, however, they

---

* *Plecotus auritus*, Geoff.

† *Vespertilio pipistrellus*, Gmel.

‡ The habits, however, of bats, in this respect, vary a little in
different places. Mr. Bell has had Natterer's bat, the whiskered
bat, and the long-eared bat, all brought him from a chalk cavern
at Chiselhurst, where they were found at the bottom of a shaft
seventy feet in depth.—The barbastelle, (*Barbastellus daubentonii*,
Bell) was taken in the same place.

left the mother, and remained by themselves in the corners of the cage, in which all were put. The old ones would not feed, and, on the night of the 16th, contrived to escape from their confinement, leaving their young behind them. Three of these young died within three or four days; but the fourth lived till the 25th, nine days after it had been deserted by its parent. During this period it appeared to be supported by a small piece of sponge saturated with milk, which was put into the cage with it; the milk being renewed once or twice a day. It clung to the sponge with great tenacity, and seemed occasionally to suck it: now and then a drop or two of milk was squeezed into its mouth.

These young must have been born some little time before they reached me, from the size which they had attained; though they were all still entirely naked, and without a vestige of hair. The largest of them measured 1 inch 11 lines from the extremity of the nose to the root of the tail, which was 11 lines more, giving the entire length as 3 inches all but 2 lines. The expansion of the wings rather exceeded 6 inches. The fore-arm measured an inch in length. The hind-feet and claws were much developed, and remarkably strong, even at this early age. The eyes were scarcely or only just opened: the teeth quite apparent, though as yet small, and not protruded far.—Of these young bats, two were males and two females.

The noctule, or great bat, withdraws earlier in the season than the common species. White says he never saw it after July. In 1829, on the last day of

August, I found one sticking to a post near some water, from which one might infer that the species was still about at that period. This individual was under size, and appeared to be the young of the year. In a previous year, on the 14th of September, I found some in a hollow tree that was evidently intended to be their winter-quarters. These were in a reduced state, and had not, as I imagined, taken food for some time.

Pipistrelle.—I have known two or three instances in which this species has been taken on wing, with the young adhering to the breast. On one occasion, an individual fell dead at the feet of a gentleman, who was out in a thunderstorm near Cambridge, immediately after a flash of lightning, as if struck by the electric fluid; a young one was clinging to the teat at the time, and still alive, apparently uninjured.*

* I have already mentioned this circumstance in a note in my edition of White's Selborne (Lett. XIV. to Daines Barrington). I reinsert it here, with the view of referring the reader to some interesting remarks, which I had not then seen, by M. Pouchet, a French naturalist, on the mode in which the female bat carries its young. These remarks are contained in *L'Institut* for 1842, p. 43, and relate to the horse-shoe bat, of which species he met with immense quantities in the subterranean vaults of an old abbey. The following is a translation of some part of his communication.

" M. Pouchet, having taken four females which had their young still clinging to their bodies, was able to observe in what manner they adhered, and resisted the sudden movements of these animals in their flight. Each female had but one young one, which adhered firmly to the mother by means of its hind-feet, in a reversed position. It embraced her indeed so closely, that, at first sight, the two animals, the forms of which were in a

Long-eared bat.—The sexes of this species would seem at certain periods to keep separate. On the 7th of September, 1825, I found a large number of individuals in the roof of a house, of which all that I examined proved to be females.

I never observed this bat abroad during the winter months, any more than the noctule. Like the last species, it appears to retire early. The pipistrelle I have seen on wing regularly every evening during a mild December.

manner confused, had a most strange appearance. When the group was carefully examined it appeared, that the young one was fixed to its mother by means of the sharp claws of its hind-feet, each of which was hooked on to the sides of the body beneath the axilla, in such a way that the belly of the young one was in contact with the abdomen of the female that carried it: its head looked backwards, and reached beyond the membrane which extends from the feet to the tail. M. Pouchet presumes that the mother, in order to allow of its suspending itself with more ease, must necessarily pass its tarsi beneath the bend of the wing of its young one.

"The adherence of these young bats to their mother was such, that the most violent shocks did not detach them. We may also imagine, that, by means of this close union, the mother, although carrying its offspring, is able to fly without inconvenience, and to go in search of food; only it would be necessary for it then to make much more strenuous exertions in order to sustain itself in the air, since it would often carry a burden, the weight of which is enormous in relation to its own, and which in the end must nearly equal it. In fact, the bats which M. Pouchet observed measured sixty *millimetres* in length from the nape to the root of the tail, and weighed twenty *grammes;* while their young ones, which appeared far from being able to abandon their mother, were already forty-five *millimetres* in length, and weighed twelve *grammes.*"

Barbastelle.—On the 17th of April, 1829, I took an individual of this rare bat in the garden at Bottisham Hall, whilst on wing.* It was a male, and flying in the middle of the day, about noon. I remember once or twice before to have seen a bat on wing in mid-day near the same place, which I judged at the time, by its manner of flight and superior size, to be different from the common sort, and which may possibly have been referable to this species. Is it a common habit with the barbastelle thus to come abroad in open day ?† The flight of this individual was slow, and near the ground.—It is worthy of note that this bat, when in hand, was perfectly free from any bad odour; wherein it differs remarkably from the noctule and the pipistrelle, both which, but especially the former, are peculiarly rank and fœtid.

* The barbastelle has since been met with in other parts of England in several instances: in Epping Forest, Mr. Doubleday is inclined to believe that it is not uncommon. See *The Zoologist*, p. 6.

† Other observers have occasionally noticed bats flying during the day. See *The Zoologist*, pp. 6 and 35. In one of the cases here referred to, "a large bat" is said to have been seen "sailing about most majestically, one sunny day in August, on Wimbledon Common." In the other, "a bat of the smaller species" was observed on wing, "on the 11th of November, at half-past three o'clock, in the wide part of the Borough (Southwark)." This last was probably the pipistrelle ; and, as the sun sets soon after four at the above season, there is nothing, perhaps, so remarkable in the appearance of this species abroad at the hour mentioned.

In a subsequent part of the above Journal, (p. 75,) other instances are mentioned both of the pipistrelle and Daubenton's bat being seen on wing in the middle of the day. The circumstance, therefore, is probably not of uncommon occurrence.

HEDGE-HOG.*

Oct. 28th, 1828.—HEDGE-HOGS are still about, and on the alert for food. I fell in with one to-day in my walks, in a sheltered part of the garden, which I was enabled to watch unobserved, and which afforded me an opportunity of seeing a little into their habits and mode of feeding. It was creeping up and down a grass walk, apparently in busy search for worms. It carried its snout very low, insinuating it among the roots of the herbage, and snuffing about under the dead leaves which lay about. After a time, it commenced scratching at a particular spot, to which it seemed directed by the scent, and drew out a very large worm from just beneath the surface of the ground. This it immediately began to devour, taking it into the mouth by one extremity, and gradually eating its way to the other; an operation which lasted some time, and was attended by an incessant action of the teeth, which grated upon one another with a peculiar noise. After the worm was all gone, as I thought, I was surprised to see the whole put out of the mouth again; and, from the appearance of the cast, I was led to believe that it had been only subjected to the action of the teeth, for the purpose of being bruised, and squeezing out the soft internal parts of the body, which alone were eaten in the first instance: the skin itself, however, was shortly retaken into the mouth, and the whole clean devoured.

From the above observation, it is probable that

* *Erinaceus europæus*, Linn.

worms form no inconsiderable part of the food of the hedge-hog, and that they are enabled to detect them by the smell, and to extract them from the ground with their snout, after the same manner that the hog uses his in searching for buried food. In the above instance no attempt was made to kill the worm before eating it; but that part of the poor creature which was still out of the mouth of the hedge-hog kept up a perpetual writhing as the nibbling of its other extremity proceeded.

COMMON SHREW.*

THE extreme voracity of the mole is well known.† The shrew, which belongs to the same natural group as the mole (the insectivorous *carnivora*), would seem to resemble it in this peculiarity, according to a statement furnished to me by my esteemed friend Mr. Selby, of Twizell. He observes in a letter received February, 1843: "What greedy gluttonous animals the shrews appear to be! One was caught

---

* *Sorex tetragonurus*, Herm.

† Mr. Bell, in his *British Quadrupeds*, quoting from Geoffroy St. Hilaire, says, "The mole does not exhibit the appetite of hunger as we find it in other animals; it amounts in it to a degree of frenzy. The animal, when under its influence, is violently agitated; it throws itself on its prey as if maddened with rage; its gluttony disorders all its faculties, and nothing seems to stand in the way of its intense voracity."

Mr. Jesse also observes, that, "as soon as the mole is caught and placed in a box, it will begin to feed with the utmost unconcern." *Gleanings in Nat. Hist.* (3rd Ser.) p. 167.

alive upon the snow here the other day, and brought
into the house, and placed in a glass box: a piece of
raw mutton was given to it, which it attacked with
the greatest voracity the moment it smelt it; and
it continued eating almost without intermission till it
had devoured the whole of it. The piece, I should
think, could not have weighed less than half or three
quarters of an ounce. When the shrew first seized
it, it shook it as a dog does a rat, and then began to
gnaw it with its sharp-pointed grinders on one side of
the mouth. It lived for a couple of days, almost
continually eating; and previous to its death, which
was very sudden, seemed in perfect health."

WEASEL.*

WHITE has observed, in his *Natural History of
Selborne*, that " weasels prey on moles, as appears
by their being sometimes caught in mole-traps."
Some years since a mole-trap was found at Bottisham
Hall, having *two* weasels in it. These animals had
both been hunting their prey in its subterranean
paths—but not, as it might be supposed, in concert:
they had come in opposite directions, and, by a
curious coincidence, must have both sprung the trap
at the same instant.

STOAT.†

MR. BELL has observed, in his *British Quadrupeds*,
that " the stoat is certainly one of the boldest ani-

* *Mustela vulgaris*, Linn.   † *Mustela erminea*, Linn.

mals of its size." As an instance of this, I may mention that I was one day sitting in my room on the ground-floor in summer with the door open, when I was surprised to see a stoat enter, and run rather rapidly about the room, snuffing about as if in search of prey. It showed not the least symptoms of alarm at finding itself in unusual quarters, and after a minute or so quietly went out again.*

Stoats and weasels, as is well known, run up trees with great facility in pursuit of birds. To-day (Sept. 3, 1827), we saw a stoat in the top branches of a tall yew-tree; but, in this instance, it appeared to be attracted by the berries, which it was eagerly devouring.

———

The circumstance of the foumart's occasionally preying upon eels is well known to the readers of Bewick's *Quadrupeds*, where there is an instance given of this fact.† Mr. Selby has observed the same predilection for this kind of food in the case of its

---

* An instance of boldness is recorded in *The Zoologist*, (p. 36,) of the weasel, which, from its similarity to the one above mentioned in the case of the stoat, I am induced to transcribe here.

"One day in June, 1842, as a lady was sitting in a room at Ilford, the windows of which opened to the ground, she was very much surprised by the appearance of a weasel (*Mustela vulgaris*), which, after trying round the window for an entrance, stood up on its hind-legs against one of the panes of glass, and remained there, notwithstanding the furious barking of a little terrier that was in the room, until the window was opened, when he started off very leisurely, but was overtaken and killed by the dog."

† Eighth edit. p. 253.

congener the stoat. The particulars which he has
sent me, connected with this discovery, are as follows.
During the course of a walk, in company with his
brother-in-law, Captain Mitford, R.N., by the margin of a rivulet in time of severe frost, and when the
ground was covered with snow, their attention was
drawn to the trace of an animal, close to the margin
of the ice, where an open current still remained un-
frozen from the swiftness of the stream. At this
spot the footmarks were very numerous, and the
animal appeared to have been frequently moving to
and fro within a circuit of a few yards. The trace,
which was supposed to be that of a stoat, was followed
to a burrow about a quarter of a mile lower down
the bank of the rivulet, the footsteps being accompanied by further marks upon the snow, which were
attributed to the prey it had secured, whatever that
might be. The entrance to the burrow was a little
larger than that of a mole-hole. Spades were immediately procured, with a view of digging the animal
out, if possible, and ascertaining what the prey was
by which the above marks were made. In following
the hole, care was taken to keep a stick constantly
in advance to mark the direction of the burrow, the
earth being soft and rather sandy. It was found to
penetrate much further and deeper than was expected; and it was necessary to dig several yards before
any approach could be made to the animal's retreat.
At length some remains were met with, which appeared to be a mass of fatty decomposing animal
matter; but which, upon closer examination, proved
to be the skin of an eel turned inside out. A little fur-

ther on, more of the same substance was observed,
as well as the remains of a water-rat, and the feathers
and bones of what seemed to have been a moor-hen.*
After a few more spadefulls had been thrown out,
they came close upon the animal, which then bolted,
but was almost instantly seized and killed by a terrier in attendance. It proved to be a male stoat of
the largest size, nearly clothed in its winter or ermine
garb ; the cheeks, shoulders, and ridge of the back
alone being interspersed with a few reddish-brown
hairs. At the end of the retreat, where the hole was
enlarged, the remains of other eels were found, and
among them one quite fresh ; which, judging from
the marks observed upon the snow, as above alluded
to, was supposed to have been caught the preceding
night. This it had begun to devour, commencing
with the head, and proceeding to eat downwards, at
the same time turning back the skin, probably by
the aid of its fore-feet; and this accounts for what
had been observed in respect of the other skins,
which were all, as already stated, turned inside out.
How the stoat succeeds in catching its slippery prey
at this inclement season, when the eel, as has been
generally supposed, is inactive, or nearly dormant, is
a problem which remains to be solved.

DOMESTIC CAT.

Cats catch and devour the common lizard,† playing with it also in the same manner they do with
mice.

* _Gallinula chloropus,_ Lath.   † _Zootoca vivipara,_ Bell.

In one instance a cat seized a more unusual prey. As we were strolling near the house, we heard the cry of some animal in distress, and, on going to the spot whence it proceeded, observed a cat crouched in the long grass of a meadow with some animal in its claws that was struggling for liberty. As we approached nearer, the cat scampered off, leaving its booty behind, which to our surprise turned out to be a large stoat, that likewise took to its legs as soon as it was released, not apparently injured. This was to me quite a new fact, that any cat, at least of the domestic kind, would attack so powerful and fierce an animal as the stoat. Had it been undisturbed, would it have devoured its prey ; or was the latter merely seized from that innate propensity to rapine and violence which characterizes the feline race ? That the stoat was conscious of being in stronger hands would appear from the cries which it uttered, and from its making no attempt to attack the cat, evidently under the influence of great alarm.

It is curious to observe what slight deviations from the course which Nature has prescribed for each species of animal are sometimes sufficient to modify, and even entirely overrule, their instincts ; at least in the case of domesticated animals, these instincts are liable to be much perverted. My cat has a kitten nearly full-grown, to which she frequently brings mice, offering them with evident symptoms of complacency, and sitting quietly by while the kitten devours them. Yet, when the

family are at meals, the old cat, who has been
accustomed to be fed from the table, is exceedingly jealous when the kitten approaches her at
such times: she is apprehensive lest the attentions
of the party should be diverted from herself to
the kitten; and, if the latter attempt to take any
of the food which she conceives intended for herself, she growls, and flies at her offspring in the
most savage manner. This has nothing to do with
any feelings of hunger, for she is often manifestly
hungry when she has caught a mouse, but which,
notwithstanding, she gives up to the kitten.

———

I observe that in summer, when my cat is prowl-
ing about the garden, and appears upon the lawn,
water-wagtails often come and alight near her, and
keep running about within a short distance of where
she is, chirping all the time incessantly. Sometimes
they suddenly take wing, but seldom fly above a few
yards, settling again almost immediately at the same
distance as before. The cat during the interval remains crouched upon the ground, eyeing the wagtail, but not attempting to spring at it, as if conscious that it was not within her reach. This continues for a longer or shorter time, when one or the
other moves further off, and the matter ends. We
can well understand that the wagtail would take care
not to approach sufficiently near the cat to be seized; but why it does not at once fly away at the
sight of its enemy, which evidently inspires it with
fear, and thus put itself beyond all danger, is not so
easily to be accounted for. It would seem like one

of the supposed cases of fascination, as it is called, of which there has been so much written by different authors.*

There can hardly be a doubt that all these cases are referable to one cause; and the only probable explanation of them, as it appears to me, is that which has been sometimes suggested, viz. that animals, under certain circumstances of great and unexpected danger, are for a time, as it were, paralyzed by fear, and so unnerved (if we may use the expression) as to be almost powerless in regard of taking the proper steps for their safety. Some have considered these cases as arising from what would be termed in man want of presence of mind. But this seems hardly taking a right view of the question, inasmuch as animals, being under the guidance of instinct, which prompts to immediate action, without any previous consideration on the part of the

* See instances mentioned in the *Journal of a Naturalist*, (3rd. edit.) p. 203 ; *Jesse's Gleanings*, (2nd. Ser.) p. 295 ; *Bell's British Quadrupeds*, p. 149.—See also an article in *The Zoologist*, (p. 293,) containing, " Notes on some peculiarities in the manners of the water-rat.

On the supposed power of fascination in serpents, see an article in the *Edinburgh New Philosophical Journal*, entitled "Fables and Prejudices regarding Serpents," by Dr. H. Schlegel, (vol. xxxvi. p. 69.) See also a note in *L'Institut*, (1842,) tom. x. p. 114. The author of this last communication, M. F. de Castelnau, asserts that he was witness to a fact in N. America, which seemed to prove to him that certain serpents really possess the power of fascination. He saw a squirrel ready to fall from the top of a tree, being fascinated by a large black snake (*Coluber constrictor*), which was at the bottom of it, coiled in a spiral, with its head elevated in the direction of the little animal.

individual, are not like human beings, who are left
to reason for themselves at every step, and who,
therefore, need continual *presence of mind*, as it is
called, not only to reason rightly, but to reason
at all.

DOG.

ANECDOTES of the dog are so extremely numerous, illustrating its instincts, and the high degree
of intelligence which it possesses above other animals, that it almost requires an apology to mention more, which, though they may be new, do not
throw any additional light on this subject.   Yet the
following may be interesting, and not out of place
in the present work, the object of which is to bring
together a few miscellaneous facts and observations
in Natural History, without attempting to refer them
to any particular principles.

A lady, living in the neighbourhood of my own
village, had some years back a favourite Scotch
terrier, which always accompanied her in her rides,
and which was also in the habit of following the
carriage to church every Sunday morning.   One
summer the lady and her family were from home
several weeks, the dog being left behind.   The
latter, however, continued to come to church by itself for several Sundays in succession, galloping off
from the house at the accustomed hour, so as to
arrive at the time of service commencing.   After
waiting in the church-yard a short time, it was
seen to return quiet and dispirited, home.   The
distance from the house to the church is three

miles, and beyond that at which the ringing of the bells could be ordinarily heard. This was probably an instance of the force of habit, assisted by some association of recollections connected with the movements of the household on that particular day of the week.

The same lady has communicated to me an anecdote, somewhat similar to the above, but more extraordinary. This related to a poodle dog belonging to a gentleman in Cheshire, which it appears was in the habit of not only going to church, but remaining quietly in the pew during service, whether his master was there or not. One Sunday the dam at the head of a lake in that neighbourhood gave way, so that the whole road was inundated. The congregation in consequence consisted of a very few, who came from some cottages close by, but nobody attended from the great house. The clergyman informed the lady, that, whilst reading the Psalms, he saw his friend, the poodle, come slowly up the aisle dripping with wet, having swam above a quarter of a mile to get to church. He went into the usual pew, and remained quietly there to the end of the service.

———

The above lady has also given me the following anecdote of a dog and cat. A little Blenheim spaniel of hers once accompanied her to the house of a relative, where it was taken into the kitchen to be fed; on which occasion two large favourite cats flew at it several times, and scratched it severely.

The spaniel was in the habit of following its mistress in her walks in the garden, and by degrees it formed a friendship with a young cat of the gardener's, which it tempted into the house,—first into the hall, and then into the kitchen,—where, on finding one of the large cats, the spaniel and its ally fell on it together, and without further provocation beat it well; they then waited for the other, which they served in the same manner, and finally drove both cats from the kitchen. The two friends continued afterwards to eat off the same plate as long as the spaniel remained with her mistress in the house.

A person of my acquaintance had a favourite spaniel that lived to be upwards of eighteen years of age. Probably not many dogs exceed that age; and I have seen some which appeared to be quite helpless and infirm, short-breathed, and without hardly a tooth in their heads, at the age of fifteen.*

SQUIRREL. †

April 23, 1822.—To-day, in a plantation of spruce-firs, we observed at the top of one of the tallest trees,

* Mr. Selby tells me, that he considers twelve or fourteen years the usual length of the dog's life. He says he has known one or two that have lived to be seventeen or eighteen, and thinks he has been told of one that had attained to twenty.

Lord Bacon, in his *History of Life and Death,* says, " The dog is but a short liver; he exceeds not the age of twenty years, and, for the most part, lives not to fourteen years." Twenty years may therefore probably be set as the extreme limit they ever reach.

† *Sciurus vulgaris,* Linn.

a large collection of moss, leaves, and slender twigs, which proved to be a squirrel's nest. It was about the size of a man's head, or larger; closed at top, with a small aperture at the side, out of which the animal sprung as we shook the tree : the nest within was spacious and softly lined. Squirrels with us are rare.

———

A friend in the North sends me the following note respecting squirrels.

" Squirrels (he says) abound in the woods at Dilstone,* where they are frequently seen gamboling among the branches, or coming on to the lawn before the house. A pair, which frequented a tree opposite the window of one of the rooms, evinced great enmity to a couple of magpies, with whom they kept up a perpetual warfare, pursuing them from branch to branch, and from tree to tree, with untiring agility. Whether this persecution arose from a natural antipathy between the combatants, or from jealousy of interference with their nests, is not known."

HARVEST-MOUSE. †

July 29, 1826.—HARVEST-MICE are common in Cambridgeshire, as might be expected, from its being so great a corn country. To-day I had brought to me the nest of one of these animals containing six young, about half-grown and well clothed with fur. It was, as White describes his, " perfectly round, and

* Near Hexham, in Northumberland. † *Mus messorius*, Shaw.

E

about the size of a cricket-ball:" but it was not
quite so firm and compact as the one he mentions;
which might have been owing to the more forward
state of the litter contained in it, and the longer
period which had elapsed since its construction. It
was composed chiefly of dry grass, with the addition.
of a few straws, and lined internally with more delicate roots and fibres. The aperture was not easily
discoverable.

LONG-TAILED FIELD-MOUSE. *

THIS species, as is well known, usually frequents
woods, fields, and gardens: in kitchen-gardens its
depredations upon the newly sown crops of peas in
the spring are often ruinous. In one instance, however, I knew an individual of this species that was
caught in a house, in the cupboard of a bed-room on
the second floor.

My brother one winter found a large collection of
beech-mast accumulated in a hole under ground, at
the foot of one of the beech-trees in the park at
Bottisham Hall. These had probably been brought
together by one of these mice, which are well known
to make hoards of food against the winter.

COMMON MOUSE.†

THE colours of the common mouse are naturally
extremely bright, and can hardly be judged of from
individuals found in houses, which contract more or

---

* *Mus sylvaticus*, Linn.   † *Mus musculus*, Linn.

less of a dingy hue from the dirt of buildings, and the nature of the recesses they frequent. To see these colours in perfection, we should have recourse to the mice found in stacks, which are often so remarkable for their bright yellow tinge, that I once thought they might prove to be a distinct species. This is due to an annulus of yellow surrounding each hair on the upper parts a little below the extreme tip, which in the domestic mouse is rarely noticeable.

It is also worthy of remark, that the mice found in stacks, except where there are many together, have comparatively little of that peculiar disagreeable smell, which is so characteristic of the common house mouse, and which is probably occasioned by its feeding upon animal and other impure substances.

WATER-VOLE. *

Aug. 31, 1829.—THE water-vole, or water-rat, as it is more commonly called here, is so extremely shy an animal, plunging into the water at the first alarm, that it is not very easy to watch its habits. To-day, however, I had an opportunity of noticing one for some time, unobserved myself. It was busied in eating the tender grasses and other plants which grew on the brink of the stream; in some cases pulling them up by the roots and devouring them entire, in others contenting itself with merely cropping the blade and uppermost shoots. In two instances I saw

* *Arvicola amphibius*, Desm.

it eat the blossoms of some species of *ranunculus*,
probably the *R. repens*, that was growing on the
bank.   During these operations, it would occasionally remain with the lower half of its body immersed in the water, having merely its head ·and
anterior extremities on land ; at other times it left
the water altogether, and strolled about the bank,
grazing as it went.   I noticed that it frequently
used its fore-feet as hands, after the manner of the
squirrel.

We have an animal frequenting the fen-ditches of
Cambridgeshire, and not very unfrequent, which
the people sometimes call the *water-mole*.   This is
nothing more than a black variety of the common
water-rat, the fur of which is sometimes of as deep
and velvety a hue as in the mole; but every gradation of tint may be found in different individuals
between this uniform rich black and the reddish-
brown which more ordinarily prevails.   There is no
other difference whatever, besides colour, between
these two kinds of water-rats, though the black has
been considered by some as a distinct species.*

On the 15th of June, 1830, I had a very large
female of the black sort brought to me, which had
been killed in the next village : it was gravid at
the time, and, on opening it, I found eight young
within, perfectly formed, and apparently quite ready
for exclusion.†

* See *Macgillivray in Wern. Mem.* vol. vi. p. 429.

† Mr. Jesse's "*sort of mole*, which partook very much of the
appearance of a rat," is, no doubt, referable to this black variety
of the water-vole. See *Gleanings*, &c. (2nd Ser.) p. 27.

I once at Ely found a small specimen of this black variety, measuring not quite five inches from the nose to the root of the tail, lying dead on the ground beneath the nest of a white owl. Like shrews, which are also often found at the foot of their nests, along with their casts, this rat appeared to have been caught and brought home by the parent owls to their young, but afterwards rejected.

BANK VOLE.*

MR. SELBY writes me word, that the bank-vole is very numerous in his garden at Twizell, and most destructive to the flower-beds. He says, it is a keen devourer of pinks and carnations; and a colony of them, which had made their way into a large turf-pit, have destroyed nearly the whole of some choice plants. He adds, they are also very fond of the Chinese chrysanthemum.

I have occasionally taken this species in Cambridgeshire in corn-stacks; but it is not near so common as the field-vole,† which abounds in our low meadows, sometimes to an extraordinary degree, as it does also in many other parts of the country.

COMMON HARE.‡

SOME years ago a hare was killed at Bottisham Hall, which, on examination, was found to be without ears. Not only was the entire auricle wanting, but even the orifice in the skin leading to the *meatus*

---

* *Arvicola pratensis,* Baillon.   † *A. agrestis,* Flem.
‡ *Lepus timidus,* Linn.

auditorius was not discernible. This appeared to have been a congenital defect, and not the result of accident.

RABBIT.*

INSTANCES of the wild rabbit have occasionally been met with at Bottisham, in which the fore-teeth had grown to so great a length as to be rendered wholly unfit for the purposes they are intended to serve. This malformation is the result of the cutting edges of these teeth not being sufficiently worn down by use, or to the degree that they are in healthy individuals, and to supply which loss the teeth themselves are provided with the power of growth. But the original cause of the evil may vary in different cases. Thus, it may be due to one pair of incisors, or one single incisor, being broken, or having fallen out; to too soft food; to a morbid and too rapid secretion of the osseous matter of the tooth, which is constantly being deposited at its root; or to some slight derangement of the under-jaw, such as, for instance, a dislocation of one of its condyles, whereby the incisors of that jaw would be thrown out of their proper position, and their cutting edges could not be brought fairly into contact with those of the opposite pair. In all these cases, either the attrition of the teeth would be checked altogether, or their growth would be over-proportioned to their abrasion by the acts of gnawing and feeding, and a preternatural elongation of that part which is above the gums

* *Lepus cuniculus,* Linn.

would immediately take place. It is obvious that
this diseased growth will be more or less rapid according to the degree of influence exerted by the
predisposing cause, and the length of time it has
operated. Perhaps, in the first stage of the malady,
its progress may be very gradual, and not much interfere with the usual habits of the animal; but the
teeth having once attained such a length, that, under
any circumstances, their edges cannot be brought to
act upon each other, their growth must be much
more rapid, and ultimately prove such an inconvenience as must often terminate in the starvation of the
sufferer.

In one rabbit, preserved in the Museum of the
Cambridge Philosophical Society, the lower pair of
incisors are so prodigiously developed as to turn
completely over the nose, and to measure in length,
from the surface of the gum to their cutting edges,
no less than two inches and one-eighth; their usual
length in ordinary individuals being only a quarter
of an inch. In another rabbit, that occurred in this
neighbourhood, but which was not preserved, both
pairs of incisors had very much exceeded their usual
length, there being also a great irregularity in their
mode of growth. The lower pair, when viewed together, assumed the shape and appearance of the letter
V, diverging from one another at the surface of the
gum, and extending in opposite directions, to the
length of nearly an inch and a half: the degree of
divergency observed in the upper pair was nearly as
great as this in the lower, and their length about the
same, but their curvature very much greater, as

indeed would necessarily result from the greater bend of that portion of the jaws in which these incisors are formed. In this instance, the portion without the gums had completed three parts of an exact circle, and their cutting edges were in close contact with the roof of the mouth.

Both the above rabbits, when taken, exhibited the appearance of having been nearly starved to death, through an inability of procuring their usual food. In the first case, life had been sustained solely by the small quantity of herbage which the animal was enabled to crop with its lips at the sides of the mouth, which appeared to have been used for that purpose. In the second instance, even this method of feeding could scarcely have been resorted to with success; the rabbit being actually unable to close its mouth, from the pressure of the lower portion of the curve, formed by the upper incisors, upon the surface of the tongue.*

In the spring of 1839, a great many young rabbits were found dead in the plantations at Bottisham Hall; which, when opened, were found to have their livers much enlarged and full of *flukes*, similar to those which cause the rot in sheep.

* The above was first published in *Loudon's Magazine of Natural History*, vol. ii. 1829, p. 134, with some further details and remarks here omitted.

HORSE.

WHEN horses are out at pasture, and resort in hot weather, as they constantly do, to the shade of some tree, they may often be observed at such times standing parallel to one another, but with their heads in opposite directions. By this arrangement, with their tails they brush the flies from one another's faces.

OX.

M. MILNE-EDWARDS has observed* that the common ox, when it lies down, after grazing, for the purpose of chewing the cud, usually reposes upon the *left* side. Any little fact of this sort has an interest, if it really admits of generalization; and I have often amused myself, when among cattle feeding in pastures, by watching them with a view to this point. But, in general, I have found the numbers lying upon the right and left side respectively so nearly equal that I can hardly attach any importance to the above statement. Still, I think I have more often observed an excess on the part of those lying upon their left side, over those on their right, than the contrary.

White, in one of his letters to Pennant, speaks of a *calculus ægagropila* which had been taken out of the stomach of a fat ox.† Professor Henslow tells me that he has got two such *calculi* or balls, that were found together in the stomach of a calf only nine

* *Elémens de Zoologie*, p. 457.
† *Nat. Hist. of Selborne*, Lett. XXXV to Pennant.

months old. The larger of the two consists chiefly
of hair, and is not perfectly spherical, being flattened
at the poles: it measures four inches and three quarters in diameter, and weighs seven ounces, three hundred and eighty-two grains. The smaller one is made
up chiefly of fragments of straw, and is spherical,
but with a very uneven surface. This last measures
two inches and a half in diameter, and weighs one
ounce, one hundred and twenty-nine grains.

SHEEP.

June 29, 1826.—OWING to the hot weather that
has prevailed of late, the cattle and flocks are dread-
fully annoyed by flies, which will not suffer them to
have a moment's rest. The cool of the evening is
the only time in which they can feed in peace. The
sheep which have been lately shorn, and whose skins
are in consequence susceptible of the slightest irritation, are in particular terribly worried by the great
horse-fly,* which eats into their flesh in the most un-
merciful manner. Hence this species does not confine its attacks to horses and the larger cattle. In
the present instance, we observed them sticking, four
or five together, to those parts of the sheep's backs
which had been just touched by the shears and
raised to a slight sore.

March 25, 1828. — A shepherd brought me to-day
for examination some minute parasitical insects with

* *Tabanus bovinus*, Linn.

which his flock are at present much infested, but
which he never observed till lately. He at first took
them for the young of the common sheep-tick ;* but,
on subjecting them to the microscope, it was at once
evident that they were true lice. It is worthy of
note, that, according to his statement, these insects
are only found upon the *long-woolled* sheep, which
are so plagued and irritated by their constant tick-
ling, that they are incessantly rubbing themselves
against posts, hurdles, and anything near, to allay
the uneasiness they suffer. It is to be remarked also,
that last winter was a very mild one; and this seems
to accord with White's observation, that sheep are
more teased in this way in such seasons.† In some
instances these lice abound in such quantities, and
cause such an intolerable itching, as to render the
sheep almost frantic ; and it is noticed at such times,
that, when rubbed on the back, they immediately turn
round and lick the hand, which gives them this momentary relief. This last circumstance may be noticed also, the shepherd tells me, in sheep infected
with the scab.

These lice generally keep close down at the roots
of the wool. They are small, being scarcely three
quarters of a line in length: the head and thorax

* *Melophagus ovinus,* Latr.—*Curtis's British Entomology,* vol.
iii. pl. 142.

† " The sheep on the downs this winter (1769) are very ragged,
and their coats much torn ; the shepherds say, they tear their
fleeces with their own mouths and horns, and that they are always
in that way in mild, wet winters, being teased and tickled with
a kind of lice." *Naturalist's Calendar,* by Aikin, p. 92.

are of a deep ochraceous yellow, approaching to ferruginous; the abdomen pale yellow, with the edges of the segments deep brown; their general form is oblong.* Adhering to many of the hairs of wool near the roots, we found a great quantity of the nits, or eggs, of these parasites. These were dirty white, of an oval-oblong form, about three-eighths of a line in length, and between two and three times as long as broad. They were applied laterally to the hairs on which they were fastened, and were furnished at one extremity with a short neck, having apparently an aperture resembling the mouth of a phial. This aperture was in all cases directed *from* the root of the wool.

Ewes that have lost their lambs, often, it is well known, have other lambs, which have lost their mothers, given them to bring up. I was much struck, in one instance, with the fact of a ewe refusing to suckle a strange lamb, except so long as the shepherd stood by with his dog; the presence of both was necessary to over-awe the poor sheep, as it were, into obedience.

The sheep-paths in a pasture, made by the animals constantly passing backwards and forwards one

* The insect above noticed is the *Trichodectes sphærocephalus* of Nitzsch, and is described and figured by Mr. Denny, in his *Monographia Anoplurorum Britanniæ*, p. 193, pl. xvii. fig. 4. Perhaps it may be uncommon in this country, or, what is more likely, has been overlooked, as Mr. Denny states his never having seen any specimen except the one which I forwarded to him for examination.

before the other, are of course bare of herbage; but I observe in most cases that the grass immediately at the sides of the path grows more luxuriantly there than elsewhere, and betrays itself at a distance by its bright green colour, forming a line completely across the meadow. This, I presume, is owing to the additional manure which it receives from the droppings of the sheep as they pass to and fro.

GRAMPUS.*

DURING a short visit at Thorney, in the Isle of Ely, in August, 1843, I was obligingly shown by a gentleman of that place an entire skeleton of some large cetaceous animal, undoubtedly a grampus, which had been found a few years back, about four miles from Thorney, in the moor, three feet below the surface of the ground. The skeleton was very perfect, and eighteen or twenty feet long. The teeth were ten in number on each side of the jaw, above and below, conical, and slightly curved: those below were rather blunted by use; those above somewhat smaller than the others, and more pointed. I was informed that three or four of these skeletons had been found at different times in the fens of that neighbourhood. This is unequivocal evidence of the sea having formerly come over the whole of that district, as indeed is sufficiently testified by other evidence.

* *Phocæna orca*, F. Cuv.

OBSERVATIONS ON BIRDS.

NOTES OF BIRDS.

Birds which are silent during the winter, as most are, appear to acquire their notes in the spring by degrees. At first their song is very weak and imperfect; and to hear them labouring at it, and only managing to get a part out, conveys the idea of some physical impediment, which for a while they are unable to surmount. As the temperature of the season advances, their system receives a corresponding stimulus, and their song becomes louder and more lengthened.* This may be particularly noticed in the chaffinch, and those birds whose song is generally made up of a definite number of notes. I have also observed it in the ring-dove, whose cooing note with us, in the height of the summer, is invariably repeated five times to complete the usual call; but in January and February, when these birds are only induced perhaps by a mild day just to try

---

* There is some allusion to this circumstance in *White's Selborne*. See his Fifth letter to D. Barrington, where I have previously given the substance of the above remark in a note, in my edition of that work.

their powers, I have sometimes heard them as if
obliged to stop after the second or third coo.

Birds also appear to lose their song in the same
gradual way in which they first acquire it. This has
been often remarked in the case of the cuckoo,
which towards the end of June is sometimes only
master of the first syllable of its call.*

Birds in general are regular to their appointed
times for commencing and ceasing song; but occasionally individuals may be heard out of season.
I do not allude here to those species which re-
assume their notes in the autumn, and even during
the winter, if the weather be mild; but more particularly to such as occasionally protract their
summer song beyond the usual period. Thus the
chaffinch, the willow warbler, and the nightingale,
all stop generally about, or shortly after, midsummer;
the nightingale, indeed, much earlier than this; but
in some years I have heard certain individuals of
each of these three species persevering through a
great part of July, and the willow wren even through
August and September. These are probably cases

* According to the old lines quoted in "Observations on some
Passages in White's Natural History of Selborne," appended to
the edition of 1813, vol. ii. p. 312.
    " Use maketh maistry, this hath been said alway,
      But all is not alway, as all men do say,
      In Aprill, the koocoo can sing her song by rote ;
      In June, of tune, she cannot sing a note :
      At first, koo-coo, koo-coo, sing still can she do,
      At last, kooke, kooke, kooke, six kookes, to one koo !"

in which some accident has occurred to break in upon the usual habits of the species; the nest perhaps destroyed, whereby the season of nidification has been prolonged; or they are cocks whose mates have been killed, and who have been unsuccessful in procuring others. When I hear a nightingale in July continuing its plaintive song so long after all its companions have ceased, with the thought that it may be due to the destruction of its first brood, such misfortune stimulating it to fresh exertions, I cannot help calling to mind those well-known lines of Virgil, which, under such circumstances, are so strictly applicable.

> " Qualis populea mœrens Philomela sub umbra
> Amissos queritur fetus ; quos durus arator
> Observans nido implumes detraxit : at illa
> Flet noctem, ramoque sedens miserabile carmen
> Integrat, et mœstis late loca questibus implet."
> *Georg.* iv. 511.

> " So, close in poplar shades, her children gone,
> The mother Nightingale laments alone,
> Whose nest some prying churl had found, and thence,
> By stealth, conveyed th' unfeather'd innocence :
> But she supplies the night with mournful strains,
> And melancholy music fills the plains."
> DRYDEN.

The following table shows the periods of the principal species of birds commencing, ceasing, and re-assuming song, or note, in this neighbourhood. It is for the most part similar to the table given by White.*

* In his Second letter to Daines Barrington.

It embraces, however, at one view, not only the singing-birds strictly so called, but others which White has arranged separately, from their scarcely deserving that name, though possessing "somewhat of a note." It also contains four species, viz. the cuckoo, ring-dove, stock-dove, and turtle-dove, which, though belonging to quite different orders of birds from all the rest, are most of them well known for their peculiar call or note, heard for a longer or shorter interval during the warm season. The whole are given in the order in which they are generally first hear.

	First heard.	Last heard.	Re-assumes.
1. Wren (*Troglodytes europæus*) . .	Jan. 5		
2. Redbreast (*Erythaca rubecula*) .	„ 8	June 23	Aug. 20
3. Common Bunting (*Emberiza miliaria*)	„ 11	Nov. 10	
4. Marsh Tit (*Parus palustris*) . .	„ 15	Apr. 18	Dec. 25
5. Hedge Accentor (*Accentor modularis*)	„ 17	July 26	Sep. 26
6. Cole Tit (*Parus ater*)	„ 19	Aug. 8	
7. Sky-lark (*Alauda arvensis*) . .	„ 23	Nov. 5	
8. Great Tit (*Parus major*) . . .	„ 24	May 22	„ 10
9. Missel Thrush (*Turdus viscivorus*)	„ 26	„ 27	Nov. 2
10. Song Thrush (*Turdus musicus*) .	„ 31	July 19	„ 29
11. Chaffinch (*Fringilla cœlebs*) . .	Feb. 1	„ 5	Sep. 2
12. Blackbird (*Turdus merula*) . .	„ 12	„ 16	
13. Golden-crested Wren (*Regulus cristatus*)	„ 14	June 23	„ 18
14. Yellow-hammer (*Emberiza citrinella*)	„ 15	Aug. 12	Oct. 19
15. Greenfinch (*Coccothraustes chloris*)	„ 21	„ 16	
16. Ring-dove (*Columba palumbus*) .	„ 25	Sep. 28	
17. Stock-dove (*Columba œnas*) . .	Mar. 5	„ 5	
18. Pied Wagtail (*Motacilla yarrellii*)	„ 17		
19. Linnet (*Linota cannabina*) . . .	„ 23	Aug. 9	
20. Chiff-chaff (*Sylvia hippolais*) . .	Apr. 3	Sep. 20	
21. Goldfinch (*Carduelis elegans*) . .	„ 11	Aug. 14	
22. Tit Pipit (*Anthus pratensis*) . .	„ 11	July 21	
23. Willow Warbler (*Sylvia trochilus*)	„ 15	„ 23	
24. Redstart (*Phœnicura ruticilla*) . .	„ 15	June 13	

	First heard.	Last heard.	Re-assumes.
25. Black cap (*Curruca atricapilla*). .	Apr. 16	July 27	
26. Tree Pipit (*Anthus arboreus*) . .	„ 20	„ 9	
27. Nightingale (*Philomela luscinia*) .	„ 21	June 7	
28. Swallow (*Hirundo rustica*)	„ 23	Sep. 7	
29. Whinchat (*Saxicola rubetra*) . .	„ 24	June 30	
30. Whitethroat (*Curruca cinerea*) .	„ 25	July 18	
31. Sedge Warbler (*Salicaria phragmitis*) „	25	„ 22	
32. Lesser Whitethroat (*Curruca sylviella*) „	26	„ 7	
33. Cuckoo (*Cuculus canorus*) . . .	„ 27	June 27	
34. Reed Bunting (*Emberiza schœniclus*)	„ 28		
35. Wood Warbler (*Sylvia sibilatrix*)	May 4	„ 28	
36. Pettychaps (*Curruca hortensis*) .	„ 4	July 12	
37. Turtle-dove (*Columba turtur*) . .	„ 8	„ 23	

The dates given in this table are, most of them, the *means* of a great many years' observations. They are the same as those given in the Calendar at the end of this work ; by referring to which, the exact number of years in each case, from which the mean is calculated, may be ascertained.—As the dates relating to this subject in that calendar are necessarily much scattered, it was thought that it might be useful to bring them together in this place, for the sake of those observers who take an especial interest in the notes of birds.

Remarks. — The wren and redbreast are usually considered as singing all the year. The latter, however, appears to me invariably to stop, like so many other species, after the breeding-season is over; it may be the end of June or beginning of July. It re-assumes, however, its note in about a month's time; and, from the middle of August on through the autumn, its singing is notorious, when scarcely any other species is to be heard. I do not feel sure that it is not the same with the wren, at least with the great majority of individuals; though here and there one may undoubtedly be heard at all seasons.

The bunting and the sky-lark are species, of which individuals may in like manner be heard during a great part of the year, if the weather be favourable; nevertheless many stop after the breeding-season, about the end of July. White says of the sky-lark, "sings in February, and on to October;" of the bunting, "from the end of January to July."

The hedge accentor, great titmouse, missel thrush, song thrush, chaffinch, yellow-hammer, and golden-crested wren, all these decidedly reassume their song in autumn in this neighbourhood, with more or less regularity, according to the character of the season. Some of them are not mentioned by White as doing this.

The marsh tit, so far as I have observed, stops earlier in the spring than the great tit, or almost any other species; but it will occasionally re-assume its note towards the end of the year. The cole tit, on the contrary, is heard during a great part of the summer.

White says, both of the thrush and blackbird, that they re-assume their song in autumn. The former, as already observed, is one of those species that may be constantly heard at that season, and sometimes on to the end of the year, if the weather be mild; but it is worthy of note, that the latter never re-assumes in this neighbourhood. The black-bird with us, from the time of its ceasing in July, is invariably silent till the following February or March. During a period of more than twenty years that I

have attended to the notes of birds, I never remember
to have heard it in a single instance in autumn. It
is also always later than the thrush in commencing
song, after the severe part of the winter is over : it
seems to wait for some decided increase of temperature, such as characterizes a fine mild day in the
early spring, of the gradual approach of which season
its note is a sure harbinger.

The greenfinch I have, in two different years,
heard for a few days in the month of December, but
it does not re-assume with any regularity.

The chiff-chaff, as observed by White,* "chirps
till September." The willow warbler also will occasionally prolong its note till that month, though
the greater number appear to stop about the middle
of July, if not sooner.

The whitethroat is also said by White to
"sing on till September." This, however, is certainly not the case here, where I never heard it after
July 28th, and the greater number of individuals
stop ten days sooner.

It is probable that, after allowing for errors of observation, it will still be found that the same species
of birds, in some instances, have different habits, in
respect of their periods of singing, in different localities. It is this circumstance which gives an interest
to tables of the above kind, brought together from
various stations, and compared with each other.

Thus, I have myself observed that thrushes and
blackbirds, which are seldom heard at Swaffham

* Called by him the "smallest willow-wren."

Bulbeck much after the middle of July, generally continue in song at Ely till the middle of August. For several years that I was in the habit formerly of spending a portion of the summer there, I noticed this fact. The earliest date of their stopping, that ever occurred to me in the precincts of that town, was the 10th of August; whilst some seasons I heard them both, on to the 26th of that month.

With regard to birds re-assuming in autumn, it seems a matter of great uncertainty, in most cases, even in a given locality. Some species re-assume with considerable regularity; but still at variable periods, and for very different lengths of time in different seasons. Occasionally they are heard for a few days only, before they are again silent; at other times they will continue singing, on and off at intervals, to the end of the year, according as it may happen.

White has observed, that, in the case of the redbreast, "many songsters of the autumn seem to be the young cocks of that year." * It has often occurred to me that this may be generally the case with all birds that re-assume in autumn, after having once ceased. Their singing at that season is, as above observed, very irregular; and their notes often few and desultory, as in young birds. Is it improbable that the young cocks may be stimulated by a fine autumn to try their powers before actually wanted; or that they may, oftener than old birds, feel prematurely those desires which are always the accompani-

* See Lett. XL. to Pennant.

ments of song in the spring of the year? At the same time, it is not meant that none of the individuals which are heard singing in the autumn may not be old birds.

———

In the height of the spring, the thrush may be heard throughout the day without intermission; but as the season advances towards midsummer, and after the solstice, especially if the weather be hot, its song is principally confined to the morning and evening. It is then one of the earliest and latest of songsters, commencing in the morning almost before it is light, and not desisting in the evening till long after sunset. During the first week in July, I have heard it in the evening as late as twenty minutes past nine.

The blackbird appears to be partial to damp weather, and, when near the time of ceasing, a continuance of rainy days, if not very cold, will often induce it to prolong its song beyond the usual period. During the fine and very dry weather which prevailed over a great part of June, 1844, the blackbirds hardly sung at all; but when the rain came, in the last week of that month, they resumed, and continued to be heard till after the middle of the month following. July was even hotter than June, but then there was much more wet: this it was which seemed to make the difference. I also once noticed, quite late in the summer, and when no blackbirds had been heard for some time previous, that one evening, after the occurrence of a violent thunder-storm, several were heard singing, but for that evening only.

The atmosphere at the time was calm, and the air mild, but extremely damp.

I have occasionally had the curiosity to note down the exact time at which the different species of birds are first heard on a fine summer's morning.* One of these instances was the 17th of July, 1826, the weather at the time being remarkably fine and settled.

The following is a list of all the species whose notes reached my ear, arranged in the order in which they struck up:

	A. M.	
Sky-lark (*Alauda arvensis*) commenced singing	2 hrs.	
Cock (*Gallus domesticus*) crowing . . .	„	40 min.
Thrush (*Turdus musicus*) first in song . .		
Yellow-hammer (*Emberiza citrinella*) ditto .	3 hrs.	
Swallow (*Hirundo rustica*) ditto . . .		
Blackbird (*Turdus merula*) ditto . . .	„	10 min.
Duck (*Anas boschas domest.*) quacking . .	„	15 min.
Pettychaps (*Curruca hortensis*) singing . .	„	25 min.
Ring-dove (*Columba palumbus*) cooing . .		
Rooks (*Corvus frugilegus*) cawing . . .	„	26 min.
Linnet (*Linota cannabina*) singing . .		
Bunting (*Emberiza miliaria*) ditto . . .		
House Sparrow (*Passer domesticus*) chirping .	4 hrs.	
Turkey (*Meleagris gallopavo*) gobbling . .		
Pheasant (*Phasianus colchicus*) crowing . .		

After four o'clock, from the fineness of the morning, the concert became so general in every direction, that it was useless making any further observations.

* This subject has been slightly touched upon in the *Journal of a Naturalist* (3rd Edit. p. 234), but not in detail, or with any attempt at exactness.

The second occasion of my pursuing these inquiries was on the 4th of July, 1843. The morning in this instance was very mild and calm, but damp and lightly clouded, inclining to rain. The following species were noticed:

		A. M.
Domestic Cock crowing	}	2 hrs. 3 min.
Blackbird chirping (but not singing), at the same time flying about from bush to bush . .		
Sky-lark singing	„	15 min.
Rooks, a few cawing	„	20 min.
Thrush singing	„	42 min.
Swallow ditto	„	47 min.
Ring-dove cooing	} „	50 min.
Thrushes everywhere singing		
Blackbird singing	„	52 min.
Redbreasts chirping (never sung at all) . .	} „	57 min.
Turtle-dove (*Columba turtur*) heard . .		
Yellow-hammer singing	3 hrs.	
Guinea Fowl (*Numida meleagris*) calling . .	„	3 min.
House Sparrows chirping	„	6 min.
Rooks beginning to leave their nest-trees . .	„	15 min.
N.B. The cawing had been gradually increasing since 2 hours 30 minutes.		
Wren (*Troglodytes europæus*) singing . .	„	20 min.
Spotted Flycatcher (*Muscicapa grisola*) chirping	„	25 min.

No other species were heard previously to four o'clock.

I attended a third time to this subject, on the 13th of June, in the present year (1845). The two former occasions were some days after the occurrence of the summer equinox; I purposely selected this a short time before. The morning was fine, the air very still, with a thin mist at times hanging over the meadows. The following occurrences were noted down.

	A. M.
Domestic Cock crowing	1 hr. 51 min.
Skylark singing	„ 58 min.
Ring-dove cooing	2 hrs. 2 min.
Collared Turtle (*Columba risoria*), in confinement, cooing	„ 5 min.
Duck quacking	„ 21 min.
Blackbird, one individual singing	„ 24 min.
Redbreast singing	„ 27 min.
Swallow singing	„ 28 min.
Thrush singing	„ 30 min.
Blackbirds everywhere singing	„ 32 min.
Guinea-fowl calling	„ 33 min.
Thrushes everywhere singing	„ 35 min.
Ring-doves everywhere cooing	„ 38 min.
Rooks, a few cawing	„ 40 min.
Cuckoo heard	„ 48 min.
Yellow-hammer singing	„ 53 min.
Rooks on wing, beginning to leave their nest-trees	„ 57 min.
Chaffinch singing	3 hrs. 3 min.
House Sparrows chirping	„ 10 min.
Jackdaw (*Corvus monedula*) on wing and cawing	„ 14 min.
Lesser Whitethroat (*Curruca sylviella*) singing	„ 25 min.
Turtle-dove heard	„ 31 min.

N.B. The thrushes and blackbirds, which had been singing everywhere incessantly from half past two, or shortly after, at this time became silent.

Starlings (*Sturnus vulgaris*) chattering	„ 40 min.
Stock-dove (*Columba œnas*) heard	„ 46 min.
Linnet singing	„ 55 min.
Greenfinch (*Coccothraustes chloris*) singing	4 hrs.
Wren singing	„ 16 min.
One Thrush resumed its song	„ 20 min.

No other species noticed previously to half past four, at which time I ceased observing.

F

On comparing the above tables, it will be seen
that on all three occasions the skylark was the earliest
of our song-birds, strictly so called, heard actually
singing. It commences about two o'clock; which,
in the first of the above instances, would be very
nearly two hours before sunrise.

> " Up-springs the lark,
> Shrill-voic'd and loud, 'the messenger of morn ;
> Ere yet the shadows fly, he mounted sings
> Amid the dawning clouds, and from their haunts
> Calls up the tuneful nations."

This fact is worth noting, because Dr. Jenner has
denied that the lark is entitled to the credit of this
precedency, giving it to the red-breast.* The redbreast is undoubtedly an early bird, but it is not,
usually, even the next after the skylark. In two of
the above instances, though heard chirping, it was
not heard to sing at all. This may have been accidental; but in the third instance it was not heard
till after the blackbird, and not till nearly half an
hour after the lark. The earliest species, in general, after the lark, appear to be the thrush,† the
swallow, the blackbird, and the yellow-hammer.
The blackbird I have repeatedly noted on various
occasions to commence about ten minutes after the
thrush, as in the first two of the above instances;
though in the third of these instances the blackbird

---

* *Phil. Trans.* 1824, p. 37.

† Mr. Thompson states that he has heard the thrush singing
in Ireland in June as early as a quarter past two in the morning.
Mag. of Zool. and Bot. vol. ii. p. 434.

was heard first. The yellow-hammer is remarkable for its great regularity in keeping to a given hour, which, during the height of summer, is three o'clock, a few minutes before or after. This species is followed generally by the chaffinch. The linnet, green-finch, and wren appear to be among the later birds, and are seldom heard till near four o'clock, if not after that hour, though the last is earlier sometimes than others.

The circumstance of the general silence among the blackbirds and thrushes after half past three, in the instance of the 13th of June 1845, is curious. Their singing had been so unceasing and so general for about an hour previous, that it required a nice ear to be able to distinguish the notes of other birds. It will be seen that at twenty minutes after four one thrush was heard to resume its song, but this was the only individual of either species that sung afterwards so long as the observations were continued.

Many birds appear to be more or less clamorous before going to roost. In the case of the rook this fact is notorious, and, before they are in sight, the approach of these birds to their nest-trees in the evening is signified to the observer by the united voices of the whole body " returning in long strings from the foraging of the day."

" Retiring from the downs, where all day long
    They pick'd their scanty fare, a blackening train
    Of clamorous rooks thick urge their weary flight,
    And seek the closing shelter of the grove."

They are often preceded, as White observes,* by
a flight of daws, which act the part of avant-couriers.
After their arrival at the rookery, they do not settle
to roost immediately, but continue sailing round and
round the trees, or hovering over them, till they almost darken the vault of heaven with their accumulated numbers, and deafen the observer standing beneath with their cries.

The common pheasant, as is well known, betrays
the place of his repose by his reiterated crowing.
The cock bird, for the hen appears to be nearly
mute on these occasions, springs from the ground on
to the tree selected for roosting, with a harsh scream
that continues unremitted till he has assumed his
perch; it is then softened into a more harmonious
crow, consisting of two or in some cases three notes,
which are repeated at intervals for a considerable
time.   Besides this cry, which is heard to a considerable distance, there is a weak inward noise immediately following, which sounds exactly like an
echo of the first, consisting of the same notes, only
in a different key, and uttered very softly.   To hear
this distinctly, it is necessary for the observer to be
almost immediately under the tree on which the
pheasant is perched.   Some individuals crow in a
much shriller key altogether than others: such, perhaps, are the young cocks of the year.

House sparrows assemble together in immense
bodies at the approach of sunset, more especially in
winter, when the shrubs in the garden at such times

* *Nat. Hist. of Selb.* Lett. XVII. to D. Barrington.

appear quite alive with their numbers. Thus congregated they keep up an incessant chirping, mixed with a shrill squeak occasionally uttered by certain individuals, exactly resembling the cry of some animal in distress. So great, indeed, is the deception, that I have been often drawn to the spot in the expectation of finding a hare in a trap, or that had been seized by vermin, or some occurrence of the like nature.

Blackbirds, prior to roosting, are not only clamorous, but singularly restless. They fly about from bush to bush in a hurried and agitated manner, as if endeavouring to escape some bird of prey, or otherwise apprehensive of danger. Their note at such times is a kind of twittering scream, different from what is usually heard at any other time of the day.

Other birds have no particular cry which they use at roosting-time, but, as the time draws on, they pursue their ordinary song with greater earnestness and vivacity. This is particularly the case with the song thrush and the hedge accentor. The latter was observed by one of our older naturalists * to keep up its song to a later period of the day than almost any other species.

It may be noticed also that swifts become more

* W. Turner, a physician and naturalist, who lived in the time of Henry VIII. He wrote a work in Latin, now scarce, on the principal birds mentioned in Pliny and Aristotle. Speaking of the hedge sparrow (as he terms it) he says, " *paulo ante vesperum solet impensius strepere, et omnium fere avium postrema dormitum petit.*"

active and noisy as the evening advances. White
remarks that, "just before they retire, whole groups
of them assemble high in the air, and squeak, and
shoot about with wonderful rapidity." On these
occasions, all the individuals of a town or village
seem to unite, in one large body, and, by way of
finale to their day's exertions, to wheel round and
round the adjoining buildings, all screaming together.

———

White has given a list of birds, few in number he
observes, that sing in the night.* In it he has included the nightingale, woodlark, and lesser reed-
sparrow.† Some other species, however, may occasionally be heard at such times. On the 30th of
April, 1843, I heard a hedge accentor singing after
midnight. The cuckoo may be often heard in the
night, as Montagu and others have noticed.‡ As an
unusual occurrence, it may be mentioned also that I
once heard the collared or African turtle (*Columba
risoria*, Auct.) cooing in a cage, out of doors, at 10
p. m. on a frosty night, in the first week of January.
It would seem, from this circumstance, that this
species, not in general considered very hardy, suffered
but little from the inclemency of our winters.

———

* *Nat. Hist. Selborne,* Lett. I. to D. Barrington.

† Now more commonly called the sedge-warbler (*Salicaria
phragmitis*).

‡ The skylark has been also heard singing at night. See
The Zoologist, p. 238.

BIRDS OF PASSAGE.

EVERY ornithologist in the country endeavours to
ascertain what species of birds are migratory in the
district in which he is resident; and to determine, as
far as possible, the exact times of their appearance
and disappearance. The following tables relate to
the birds of passage hitherto noticed by myself in
the vicinity of Swaffham Bulbeck, arranged somewhat in the order in which they are first seen:*

I. *Birds that spend the summer, and breed, with us.*

1. Pied Wagtail (*Motacilla yarrellii*) . . . — Seldom seen here before February or March, though in mild winters a few remain with us the year round.

2. Chiff-chaff (*Sylvia hippolais*) . . . . — First heard the end of March, or beginning of April, but not very plentiful.

3. Redstart (*Phœnicura ruticilla*) . . . . — In general, very regular in its appearance about the 12th or 13th of April.

4. Willow Warbler (*Sylvia trochilus*) . . .
5. Blackcap (*Curruca atricapilla*) . . . . — Both these also very regular: appear usually about the same time as the redstart; the blackcap, however, I once heard in March.

6. Great Plover (*Œdicnemus crepitans*) . . . — Middle of April: small flocks sometimes seen in the neighbourhood of Newmarket Heath.

---

* Similar to the tables given by White in his first letter to
Daines Barrington.

7. Swallow (*Hirundo rustica*) { Usually first seen about April 19 : stays till the middle of October.

8. Sand Martin (*H. riparia*) { This species is not found nearer than Quy Water, (three miles from hence,) where I have observed it usually about the third week in April.

9. Tree Pipit (*Anthus arboreus*) . . . .

10. Nightingale (*Philomela luscinia*) . . . .

About April 20.

11. Wryneck (*Yunx torquilla*) ·· April 22.

12. Wheatear (*Saxicola œnanthe*) . . . . { A few of these birds are seen upon Newmarket Heath, where they breed in holes on the Devil's Ditch ; but they are not very numerous, and I am uncertain at what period they arrive.

13. Whinchat (*Saxicola rubetra*).

14. Whitethroat (*Curruca cinerea*) . . .

15. Lesser Whitethroat (*C.sylviella*) . . . .

16. Sedge Warbler (*Salicaria phragmitis*) . . .

These four species appear very much together ; usually about April 25.

17. Cuckoo (*Cuculus canorus*) { Seldom heard with us before the last week in April, and sometimes not till May.

18. Yellow or Ray's Wagtail *Motacilla raii*) . . } End of April.

19. House Martin (*Hirundo urbica*) . . . { End of April or beginning of May : generally stays a few days later than the swallow.

20. Pettychaps (*Curruca hortensis*) . . . . } Beginning of May.

21. Wood Warbler (*Sylvia sylvicola*) . . . { Ditto. Not common with us, and generally heard but a very short time ; some years not at all.

22. Turtle-dove (*Columba tur-tur*) . . . . Generally the first week in May, but sometimes in April. Remains till September, if not longer.

23. Common Sandpiper (*Totanus hypoleucos*) . . April or May, but not common.

24. Swift (*Cypselus apus*) . Never seen here before the 6th of May, and sometimes not till the third week : departs early in August, if not sooner.

25. Spotted Flycatcher (*Muscicapa grisola*) . . About May 16 : monotonous chirp heard till late in the summer.

26. Quail (*Coturnix vulgaris*) . 27. Landrail (*Crex pratensis*) . These species are seldom heard with us before the middle of May : how much earlier they arrive I am not certain. Some quails remain the winter.

28. Hobby (*Falco subbuteo*) . 29. Red-backed Shrike (*Lanius collurio*) . . . 30. Goatsucker (*Caprimulgus europæus*) . . . Met with occasionally during the summer months, but not sufficiently plentiful to determine the exact date of their arrival.

31. Ruff (*Machetes pugnax*) . 32. Black Tern (*Sterna nigra*) Resort occasionally to our fens to breed, but only in very wet summers; are getting scarcer every year; ruffs especially very seldom seen now.

II. *Birds of double passage.*

33. Ring-ouzel (*Turdus torquatus*) . . . A few years back some of these birds were observed for two or three seasons in succession, each spring and autumn, in the adjoining parish of Swaffham Prior; but I have not heard of their having been seen there lately.

34. Dotterel (*Charadrius mori-nellus*). . . . { Makes its spring passage between the 20th and 30th of April: returns about the third week in September.

III. *Birds that spend the winter, or a portion of the winter, with us.*

35. Golden Plover (*Charadrius pluvialis*) . . . { Arrives the middle or end of September: stays till March or April.

36. Snipe (*Scolopax gallinago*) { Middle of September; but many remain the whole year, and breed in the fens.

37. Jack Snipe (*S. gallinula*) { End of September: never stays to breed.

38. Short-eared Owl (*Otus brachyotos*) . . . { September or October; plentiful some years in the fens.

39. Bean Goose (*Anser segetum*) . . . . { Earliest flocks seen the beginning of October; return northwards in March: much less plentiful than formerly.

40. White-fronted Goose (*A. albifrons*) . . .

41. Grey Wagtail (*Motacilla boarula*) . . . } Beginning of October.

42. Wild Duck (*Anas boschas*) .. Middle of October.

43. Woodcock (*Scolopax rusticola*) . . . . { Middle or end of October: stays some seasons till near the end of March.

44. Hooded Crow (*Corvus cornix*) . . . . { October or November: last seen about the second week in March.

45. Fieldfare (*Turdus pilaris*) .
46. Redwing (*T. iliacus*). . { Rarely seen before the end of October. Sometimes fieldfares remain very late in the spring; occasionally till near the end of April.

47. Teal (*Anas crecca*) . . { Has occurred the first week in October, but generally later. Not seen every year.

48. Hooper (*Cygnus ferus*) . } Seen only in very hard win-
49. Bewick's Swan (*C. bewickii*) } ters.

50. Pintail Duck (*Anas acuta*)
51. Wigeon (*A. penelope*) .
52. Pochard (*Fuligula ferina*) Visit our streams and ditches
53. Scaup Duck (*F. marila*) . occasionally in hard weather,
54. Tufted Duck (*F. cristata*) but not of regular occurrence.
55. Smew (*Mergus albellus*) .

56. Great Grey Shrike (*Lanius* Now and then one killed
     excubitor) . . . during the winter.

57. Brambling (*Fringilla mon-* Appear at rather uncertain
     tifringilla) . . . intervals during the winter, and
58. Siskin (*Carduelis spinus*) . not every year. Siskins most
59. Twite (*Linota montium*) . often seen in January and Feb-
60. Lesser Redpole (*L. linaria*) ruary.

EQUAL DISTRIBUTION OF BIRDS.

THERE is, in this neighbourhood, an annual destruction of rooks and sparrows to a great extent
every year; yet I observe, as others have observed
before in similar cases,* that no apparent diminution
of their numbers takes place. The rookery at Bottisham Hall is very large, and has existed from time
immemorial; it appears to be the head-quarters,
whence many small colonies, which are now established wherever there are a few tall trees in the surrounding neighbourhood, originally emanated. And
to these head-quarters the inhabitants of the small
scattered colonies appear to return in winter, flocking
with the general mass, till the approach of the breeding season. The number of young birds in this
rookery, either shot, or taken unfledged from the

* See *Journal of a Naturalist* (3rd edit.), p. 183.

nest, has amounted in some years to nearly a hundred dozens; to say nothing of the old birds which are occasionally destroyed at all seasons of the year. Yet I feel satisfied that the general aggregate of the nests in spring, as also of the individuals forming the immense flocks we see in autumn, is much the same as it was thirty years back.—So too with the sparrows, which abound in such multitudes, to the great annoyance of the farmer; they are not, to my thinking, more or less abundant than they were formerly, though here, as in other places, the parish officers give rewards for their destruction every year.

My idea in regard of this matter is, that species, which, like the rook and sparrow, are generally dispersed over the kingdom, have a tendency to equalize their numbers throughout the different parts of it; deficiencies from accident, or any other cause, in one locality, being made up by supplies from another, in which there happens to be an excess. It seems also not improbable, that, without a diminution of numbers in any one place to induce strangers to settle there, the numbers in another are not allowed to increase beyond a certain limit, in consequence of the law which impels old birds to drive away their offspring, as soon as they are sufficiently matured to shift for themselves. If there be convenient space for them in the neighbourhood, without interfering with the old birds, they may be suffered to remain; but otherwise they are forced to go elsewhere; and if there is no colony near, allowing of increase, to which they can attach themselves, they found a new colony, wherever circumstances may be favourable to their so doing. In some instances, they may have to traverse long distances before they can effect this object; or they may be compelled to leave the country altogether, and resort to another; and perhaps this supposition may serve to explain the circumstance of their sudden appearance in some localities, where they had not been previously observed.* The well-known fact, too, of old birds driving away their young may, in the case of certain rare species, account for their being young or immature individuals, which alone occur, perhaps, at irregular intervals, in a given country. These individuals are such as have wandered to near the extreme geographical limits assigned to their species. If, from a general excess of numbers in other places, they are compelled to travel so far from central quarters as to overstep these limits, existence is with difficulty maintained, or may be no longer possible. They then perish; and it may be in this way that the numbers of some species are constantly held in check, or at least reduced, in certain seasons, when particular circumstances have led to an undue increase.

It seems further deserving of inquiry, whether the flocking of birds in some instances, as well as the

* An instance of this kind is mentioned in the volume of *Reports on the Progress of Zoology and Botany*, lately published by the Ray Society. "In April, 1838, a flight of rooks" is said to have "entered into the city of Danzig, and settling upon all the larger trees, in gardens as well as in the most crowded streets, built their nests there and brooded."—*Prof. Wagner's Report on Birds*, p. 71.

partial flittings of individuals from one part of the
country to another, may not arise from the desire,
perhaps the necessity, of seeking some new abode.
White has observed, that " there are, doubtless,
many home internal migrations within this kingdom
that want to be better understood"; * and though
the instance he adduces of the vast flocks of hen
chaffinches that appear at Selborne in the winter is
probably quite independent of the above cause, his
remark is not the less deserving of consideration in
reference to the subject we are speaking of.　Flocks
of individuals, of several species, may occasionally
be seen, at all seasons, wending their way steadily in
a direct line, as if under the influence of some common impulse.　There are also flittings which appear
to have nothing to do with the ordinary migrations,
inasmuch as they occur in species which do not, so
far as we are aware, migrate.　It would be an
interesting record if naturalists, resident in the
country, would enter in their journals any facts of
this kind that might come under their observation,
as also any cases of new species taking up a permanent abode in their neighbourhood, † or of old
ones experiencing a marked increase, or falling off,
of their numbers.

　　There is one circumstance, not wholly uncon-

　　* Eighth letter to Pennant.

　　† One such instance in the case of the turtle-dove, a migratory species however, I have mentioned further on in this work.
This species was not observed in the neighbourhood of Bottisham
and Swaffham Bulbeck before the year 1823.　Since then they
have visited it regularly each season, and in increasing numbers.

nected with this last point, which seems to have been frequently noticed by English naturalists; and that is, the fact of certain species, which flock regularly in winter, being observed at such seasons in larger numbers than can well be supposed to have been bred in one neighbourhood. This is White's observation in regard to the flocks of hen chaffinches above spoken of. The same idea has been entertained by others, in the case of the large flocks of greenfinches, linnets, buntings, and some other small birds, which are everywhere so abundant in this country in winter. It is clear that such flocks cannot, in all places at once, exceed in numbers the accumulated individuals of the respective localities in which they occur; unless we suppose, as suggested by some naturalists, that they receive their accessions from abroad. * This, indeed, may possibly be the case : and it may further happen, that many of the strangers thus coming in amongst them may be induced to remain on in the spring, where there is an opening to allow of their gaining a settlement; supplying in this manner another means, in addition to those before suggested, by which the original numbers in any particular spot, where they were deficient, may be recruited.

NESTS OF BIRDS.

I observe that birds sometimes begin the shell of their nest, or the outer part of the fabric formed

* See *White's Selborne*, Lett. XIII. to Penn. *note* in my Edition.

of the coarser materials, before it is wanted, and let it remain in this state for some days, and then finish it on a sudden. In other cases I have known the nest entirely completed, and afterwards a fortnight or more elapse before laying commenced. This last circumstance I have observed occasionally in the chaffinch; and had been led to suppose in such instances that the nest from some cause had been deserted. I passed it day after day, and found it exactly in the same state, without any eggs in it. But it proved, in the end, to be only a delay, from the circumstance of the nest not being needed sooner. After an interval of longer or shorter duration, the eggs were duly laid, and the brood reared. I have noticed the same habit in the tree creeper.

BIRDS APPEARING FAT IN WINTER.

THE circumstance of birds puffing themselves out, so as to appear plump, in severe weather, has been often noticed.

> " The redbreast, ruffled up by winter's cold
> Into a feathery bunch,"

has not escaped the observation of the poet. But if this is what White alludes to, when he speaks of "birds growing fatter in moderate frosts," I conceive he is wrong as to the fact, to say nothing of the way in which he attempts to explain it. His remark, indeed, in the first instance, is in reference to "long-billed birds" (the *Grallatores* I presume he means); with regard to which it may be true, though I am not aware of it. But then he adds, "the case is just the same with blackbirds," &c.* Here I should be disposed to doubt the fact altogether of these birds being fatter in frosts than at other times. They may appear to be so; but it is appearance only, and, as I imagine, simply an expedient to which they resort in order to keep up better the temperature of their bodies in cold weather. By ruffling up their feathers, they admit the air between them, which is a worse conductor of heat than the feathers themselves, at the same time that it removes further from their bodies the conducting surface by which their natural heat would be carried off; just as a loose cloak about the person is warmer than one which fits closely to it. The conducting surface itself also is less effective when the feathers partially stand up than when they are smoothed down; though the feathers radiate more freely under the former condition than under the latter. This is according to the known laws, by which, *cæteris paribus*, bodies radiate and conduct heat according to the nature of their surfaces.

MORTALITY AMONG BIRDS.

It has been thought by some a matter of surprise, that we so seldom meet with dead birds,† notwithstanding the immense multitudes that must die every year. It is said that man destroys few, comparatively speaking; and it is asked what becomes of the re-

* Letter V. to Daines Barrington.
† *Jesse's Gleanings*, Second Series, p. 296.

mainder. This inquiry leads to several other inquiries connected with the statistics of birds, which it would be most interesting to have answered; but which, in the present state of our knowledge, it would be difficult to clear up. More especially it would be important to determine at what period, as regards both the age of the individual, and the season of the year, as well as how and in what exact manner, the mortality among birds takes place. This mortality must, in the instance of some of the smaller species, be very great. The numbers reared each year seem to bring no proportionate accession to the general number of individuals the year following.

Does then the mortality fall principally upon the old birds or the young, and during winter or during summer? White has observed with respect to house martins, that " they must undergo vast devastations somehow, and somewhere; for the birds that return bear no manner of proportion to the birds that retire."* From this one might suppose, that, in the case of this species, it was during their sojourn in a foreign country in winter that their numbers became thinned, and that the deaths were in proportion to the increase they had received here during the breeding season; for about the same numbers seem to return each year. It would follow also that it was not by the destruction which fell upon the very young birds, that these numbers were kept within the assigned limits, but by that of the adult. I think, however, it may fairly be questioned, whether,

* Letter XVI. to Daines Barrington.

because martins return to this country in smaller numbers than they quit it, it necessarily follows that the overplus have all died abroad during the winter. The excess of numbers at the time of migration in autumn is made up of the young birds of that year. Now, though it seems well ascertained, that *old* birds, which have bred in this country, generally return to the same breeding-places each season, I am not aware of any facts that prove that all the *young* birds return with their parents. Is it improbable that, instead of doing this, many may disperse themselves about in various directions, when the spring arrives, and assist in filling up gaps, wherever, from accident or other cause, the numbers in any place may be deficient? It may be in this way, partly, that there is kept up, in the case of some species, that equal balancing of numbers in different countries, to which we have already alluded in a former place.*

If it be indeed true, that, with regard to martins, a vast devastation of individuals (as White supposes) takes place somewhere in other countries during winter, and if the same could be shown of our summer migrants generally, analogy would lead us to expect that our winter migrants would receive the check to their undue increase, in like manner, during winter, in *this* country. And according to this view these migrants must visit this country in autumn in much *larger* numbers than those in which they quit it in the spring. But I doubt if we have any ob-

* See back to p. 108.

servations on record fully to establish this point in
all cases, however it may hold in some. After all,
this leaves the original question, as it regards those
species which remain with us the whole year, un-
touched; and these constitute a great proportion of
the whole. How then, it may be asked, is it with
respect to them?

For my part, I should be inclined to think that, in
most instances, the mortality would not be confined
mainly to the old birds or to the young, but divided
between them; and that it was not, on the whole,
more general during one season than the other.
Young birds, that had but lately quitted the nest,
would, during summer and autumn, most readily
fall a prey to rapacious animals; * while the old,
from their enfeebled powers, would perhaps be
less able to contend with the severities of winter.
And a certain proportion of these also must furnish food for the *Raptores*, at least *during winter*.
Probably then, from these causes combined, but
few individuals attain to old age, or fail of meeting,
sooner or later, with an untimely end. The law
by which animals prey upon each other, seems to
be an express provision of Nature, as a means of ter-
minating their existence, without subjecting them
to that amount of previous suffering and priva-

---

* Thus White, speaking of a pair of sparrow-hawks, in the
month of July, says, that they "had been observed to make sad
havoc for some days among the new-flown swallows and martins;
which, being but lately out of their nests, had not acquired those
powers and command of wing that enable them, when more ma-
ture, to set such enemies at defiance."—*Lett.* XLIII. *to Pennant.*

tion, which would necessarily, in their case, attend death from old age.* In those species which are not the prey of others, but which are themselves the devourers, the number of individuals of each sort is generally small; and, doubtless, there exists some other provision in their case, causing death at an appointed age, before life has been protracted to its utmost limit. There seems to be a reason, then, at least in the case of the smaller birds, why we should not oftener meet with dead individuals. A considerable proportion undergo a violent death: of those which die in other ways, probably many are carried off by small predaceous quadrupeds almost as soon as they fall to the ground; while the rest are left for the scavenger insects to devour or bury, and it is surprising in how short a time, during the warm months of summer, some of the burying beetles will accomplish the humble task allotted them by Providence.†

* This subject has been before alluded to in the Introduction to this work. See p. 24, *note*.

† I have repeatedly, in spring, placed dead birds on the ground in different spots, and found them so completely buried by the *Necrophorus vespillo*, and other allied species, within twenty-four hours after, or nearly, that I have sometimes had difficulty in finding them again.

NOTES ON PARTICULAR SPECIES OF BIRDS.

KESTREL.*

April 22, 1823.—THIS morning we found a kestrel's nest in one of the highest branches of a tall spruce-fir, containing five eggs, all of equal size, and coloured alike † : four is the more usual number.

April 27, 1831.—Another nest of the kestrel was found to-day, with one egg in it. The old bird was shot as it flew off, and proved to be the male.

This species of hawk feeds much on insects occasionally, as appears by their casts and fæces, which often consist entirely of the remains of beetles.‡

* *Falco tinnunculus*, Linn.

† Mr. Yarrell observes that, when there are five eggs, "the fifth has been known to weigh several grains less than any of those previously deposited, and it has also less colouring matter spread over the shell than the others ; both effects probably occasioned by the temporary constitutional exhaustion the bird has sustained in her previous efforts."

‡ This fact has been noticed by others.—See *Ann. and Mag. of Nat. Hist.*, vol. xiii. p. 93.

HONEY BUZZARD. *

Sept. 23, 1826.—A honey buzzard was trapped at Bottisham to-day at a wasp's-nest. The keeper had observed it there previously several different times, tearing the nest in pieces, and on one occasion actually carrying away large pieces of the comb in its bill and talons.† The plumage of this individual seemed to indicate that it was a young bird of the year. When the mouth was opened, a very perceptible smell of honey proceeded from the gape. This species of buzzard has been killed in this neighbourhood in three or four instances.

MARSH HARRIER. ‡

THIS species of harrier is very plentiful in the low grounds of Cambridgeshire, making its nest on the ground among tall grass and rushes. I have had the newly-fledged young brought me from Burwell Fen, the second week in May. These have uniformly wanted the yellow patch on the crown of the head, so conspicuous in the adult bird. It is rather a variable species in respect of plumage, being some-

---

* *Pernis apivorus*, Gould.

† Mr. Selby has also trapped this species at a wasp's nest, in Northumberland. See *Proceed. of the Berwickshire Naturalists' Club* for 1836, p. 109, where he has given a detailed account of the capture.

‡ *Circus rufus*, Selby.

times found with the lower half of the abdomen entirely white, and the other parts of the body here
and there spotted with that colour.   This, and the
common buzzard, bear indiscriminately, in these
parts, the provincial name of *Puttock*.

LONG-EARED OWL. *

April 16, 1827.—THIS morning we found the nest
of a long-eared owl in an old ivy-tree much tenanted by jackdaws, one of whose holes had been mono-
polized for the purpose.   This confirms what authors
state as to this species of owl not making any nest
of its own, but availing itself of that of some other
bird.   The nest contained two eggs, which had not
yet been incubated.   The long-eared owl is occasionally, but not very commonly, met with at Bottisham.   It seems partial to the gloom of thick fir
plantations.

SHORT-EARED OWL. †

THIS species of owl is not uncommon throughout
the low grounds of Cambridgeshire, where it makes
its first appearance towards the latter end of September.   I have been informed that in the fens, in the
neighbourhood of Littleport, it is sometimes found
in astonishing plenty, particularly after those seasons which have been most productive of field-mice,
which appear to be its favourite food and a great

---

* *Otus vulgaris,* Flem.   † *Otus brachyotos,* Flem.

object of attraction. In those districts they are
known by the name of *Norway Owl.* Their usual
haunts are fields of cole-seed and turnips ; in which
situations they may often be put up one after another, to the number of fifty or more : but they are
never observed in stubbles, or amongst trees during
the day; though they resort to these last to roost in
at night, and at such times seem much attached to
plantations of spruce-firs.

WHITE OWL.*

I formerly spent a portion of the summer, for
several years in succession, in a house at Ely, the
eaves of which were tenanted by a pair of white
owls, that bred there regularly every season ; and
I had an opportunity of testing the accuracy
of what White has said respecting the habits of
this species. I never observed them bring anything
home to their young except mice and small rats.
Amongst the former, shrews were not unfrequent ;
but these, though captured, did not appear to be
in general devoured, as they might constantly be
found entire on the ground beneath the nest, having been cast out, along with the usual rejected
pellets of bones and fur. I once noticed amongst
these rejectamenta a specimen of the water-shrew ;†
and on another occasion a half-grown individual of
the black variety of the water-rat mentioned in a
former part of this work. ‡   I never in any instance

* *Strix flammea*, Linn.   † *Sorex fodiens*, Gmel.
‡ See p. 76.

heard this species of owl hoot, (agreeably with what
White observes,) though I have attended to its habits
for a very long period. I mention this, because some
naturalists have denied the accuracy of White's statement on this point.

————

White has mentioned a tame brown owl, with
which he was acquainted.* A friend of mine has
sent me the following particulars respecting a tame
white one, which was taken, when young, from a
nest in the woods at Dilstone, near Hexham in
Northumberland, and given by a lady to her
children, who brought it up. Great pains appear
to have been taken to domesticate this owl, in
consequence of which it became very familiar.
In imitation of its own call, it received the name
of *Keevie*, to which it would readily answer when
within hearing, following the sound from whatever
part of the premises it might happen to be in. Its
usual place of repose during the day was under the
branches of an old Scotch fir, which grew down a
steep inaccessible bank, where it would sit apparently asleep, but sufficiently awake to endeavour
to attract the notice of any one who passed, by its
usual cry of *keevie, keevie.* If the passenger stopped and answered it, it immediately scrambled up
the boughs of the fir, till it brought itself to a level
with the walk above, in hopes of being fed; but
if he went on again, unheeding its solicitations, it

* *Nat. Hist. of Selborne,* Lett. XI. to Pennant.

returned to its former place, and resumed its slumbers. One of the most striking peculiarities in this tame owl is said to have been its fondness for music. It would often come into the drawing-room of an evening, on the shoulder of one of the children, and, on hearing the tones of the piano, would sit with its eyes gravely fixed on the instrument, and its head on one side in an attitude of attention; when, suddenly spreading his wings, he would alight on the keys, and making a dart at the performer's fingers with its beak, would continue hopping about, as if pleased with the execution.

After a while the flights of this owl into the woods became longer, and he only returned at dusk to receive his usual supper from the person who was in the habit of feeding him, and whom he readily permitted at such times to take him up, and carry him into the house for this purpose. Bye-and-bye it was observed that he did not devour his meal in the kitchen as formerly, but fled along the passage, dragging the meat after him, till he reached the garden-door, when he flew with it to a part of the shrubbery: on being followed, it was discovered that he had brought with him a companion, who, not having courage to accompany him the whole way, remained at a respectful distance to receive his bounty. After having served his visitor in this manner, he returned to the kitchen, and leisurely devoured his own portion. This practice was continued for some months, till at length one evening he was missed, and nowhere to be found: his companion, it is said, continued to visit the spot alone for several weeks uttering doleful cries, but could never be persuaded to come nearer to be fed. It proved, in the end, that the favourite had been killed; and its stuffed skin was one day recognized, alas! in a woodman's hut, by the children, who had so assiduously nurtured it and brought it up.

TAWNY OWL.*

THIS species appears to be an earlier breeder than any of our other native owls. In some seasons I have known the young hatched by the end of March.

RED-BACKED SHRIKE.†

I HAVE occasionally examined the stomach of this species, but never found it to contain anything beyond the remains of coleopterous insects. From the statements of other observers, however, it would seem that it does undoubtedly sometimes prey on small birds.

MISSEL-THRUSH.‡

WHITE observes that the missel-thrush " is called in Hampshire and Sussex the *storm-cock,* because its song is supposed to forebode windy wet weather." The fact is, that it sings very early in the season, sometimes commencing with the new year: but it

* *Syrnium aluco,* Jen.   † *Lanius collurio,* Linn.
‡ *Turdus viscivorus,* Linn.

does not generally sing, except the weather be mild; which is seldom the case in January and February, without more or less wind and wet accompanying. The missel-thrush is plentiful about us; and its notes, which are rather loud and powerful, come grateful to the ear at a time when there is but little vocal music to be heard in the woods. Many persons mistake it for the common thrush, but its song has not the variety which belongs to that of this last species. It mostly frequents the tops of the highest trees, where it sits singing on a naked bough for a considerable time without intermission,* and like the common thrush seems to delight in one particular spot. Besides its regular song, it has a kind of hissing jarring note, which it uses as a menace during the breeding season.

REDWING† AND FIELDFARE.‡

Feb. 20, 1827.—THE stomach of a redwing shot this morning was found to contain the remains of coleopterous insects, earwigs, and spiders; some of these last were quite perfect and entire: there were also a few shells of the *Helix hispida.*—The stomach of another individual, shot at the same time, was almost entirely filled with the above shells, and contained little else.

The stomach of a fieldfare, killed at the same time

---

* This peculiarity in the missel-thrush, of its song being carried on continuously for a longer time than in most birds, has been noticed by another observer. See *The Zoologist,* p. 492.

† *Turdus iliacus,* Linn.   ‡ *Turdus pilaris,* Linn.

as the two redwings, was crammed tight with the half-digested remains of berries,—chiefly haws.

The above facts seem to indicate a difference in the food of these two species of birds, the redwing inclining more to an insect, or animal diet, than the fieldfare ; and if so, this difference may serve in some measure to account for a remark of White's, that "when birds come to suffer by severe frost, the first that fail and die are the redwing fieldfares."* I have myself occasionally noticed the same circumstance, and many times picked up redwings during hard weather in a dead or starving condition, whilst I never met with a fieldfare in the same state. Insects and shells must be obtained with considerable diffi-culty as the severity of the season increases, whereas of berries there is seldom a deficient supply.†

SONG-THRUSH.‡

THRUSHES, as White and others have observed, live much on snails during the summer, especially in

---

* *Nat. Hist. of Selborne*, Lett. V. to D. Barrington.

† Mr. Thompson has also made some remarks on the food of these two species of birds, which appear rather to confirm the above idea. The stomachs of three redwings, opened by him, contained principally insects and shells, mixed however in two individuals with some vegetable food (chiefly bits of grass). The stomachs of three, out of six fieldfares, were found to contain vegetable matter alone (mostly grain) ; one being "filled with oats, though the weather had been mild for some time before, and when it was shot :" the other three contained insects, shells, and worms, in addition to the vegetable matter.—See *Mag. of Zool. and Bot.* vol. ii. pp. 433 and 437.   ‡ *Turdus musicus*, Linn.

dry weather.* They appear to resort to particular
spots and favourite stones for breaking the shells of
these animals. There are two or three such spots in
my garden, which are very much visited for this
purpose, and where the shelly fragments of the
Helix aspersa may be found in some seasons accumulated by handfuls. It is very amusing to watch
the thrush holding the snail in its bill, and forcibly
knocking it against a stone, in order to get at the
contents of the shell. The rapping noise which it
makes may be heard to a considerable distance, and
I have often known persons puzzled at such times to
know whence the noise proceeded. When disturbed
during the operation, I have seen the bird fly off with
its booty to another spot.

REDBREAST.†

A LADY has furnished me with the following
striking instance of maternal affection in a redbreast,
that had built in some ivy against a wall in a garden
at Whitburn, near Sunderland, in April 1839. The
bird was sitting upon four eggs, when the gardener
one day trimmed the ivy so close with his shears as

---

* And sometimes also in winter, it would seem, during frost,
when other food is scarce, according to Mr. Selby. In a letter,
dated Twizell, Feb. 12th, 1845, he observes: "We have had
severe frost, and a good deal of snow, since the commencement of
the month. The thrushes are all now near the coast, about the
roots of the thickest hedges, where they obtain a supply of *helices*:
as soon as a thaw takes place, we shall have them in the plantations." † *Erithaca rubecula*, Swains.

almost to destroy the nest; in consequence of which the eggs were precipitated to the ground. They lay there till observed by the lady shortly afterwards, who was attracted to the spot by the plaintive cries of the parent bird. It was at first thought that to restore them to the nest would prove useless. The attempt, however, was made; the eggs, which were nearly cold, were picked up, and placed back again in the nest, after it had been repaired and put together again as well as was possible. They had not been returned to their former situation five minutes, when the bird came, and again took charge of them, and in two days they were hatched; the infant brood being from that time, of course, objects of daily interest and observation. Great was the dismay of the lady, some days afterwards, at finding all the little ones upon the ground, stiff and cold, having fallen through a fracture in the patched nest, which was not sufficiently strong to keep together. She took them up, and perceiving a slight movement in one of them, carried them into the house, where, partly by the warmth of the hand, and partly by the influence of a fire to which they were held, they all gradually recovered. They were then again placed in the nest, which was further patched with a piece of drugget, fastened into the fracture through which they had fallen. They were doomed, however, to go through more trials; for it happened, some nights after, there was a heavy rain, which so completely soaked the nest, and the drugget which had been placed in it as a lining, that the young ones were found the following morning almost drowned,

and to appearance lifeless. They were again brought
to the fire, and thoroughly dried; after which they
were placed in the empty nest of another bird that
was substituted for the old one, and fixed in a
currant-bush, a few yards from the wall where the
ivy was. The young ones, which were half-fledged
when they got this wetting, still continued to receive
the attentions of their parent; and in due time they
were all safely reared, and flew away. It is stated,
that it was very curious to observe the familiarity of
the old birds during the whole course of these proceedings: they always sat close by, and never seemed
the least alarmed at the liberties taken with their
progeny.

————

The pugnacious disposition of the redbreast
towards its own kind, as well as towards other birds,
is well known. Mr. Selby sends me the following
remarkable anecdote, shewing to what an extent
their passion will sometimes carry them, and how
completely they are lost to all apprehensions of danger
while under its influence. A redbreast had for
some time taken up its abode in a hot-house, from
which it had egress at pleasure. One day, when the
gardener was in the house, another redbreast found
his way in; but he had no sooner made his appearance than he was furiously attacked by the usual
tenant, and soon shewed that he had the worst of the
combat; so severely was he treated, that he was
taken up by the gardener, and held in his hand,
where he lay struggling and panting for breath.

The victor, however, was not thus to be deterred
from further wreaking his vengeance upon the in-
truder.   He boldly flew, and alighted upon the hand
of the gardener ; and forthwith proceeded to peck the
head of his victim, and buffet him in such a manner
that he would soon have put him *hors de combat,* had
not the gardener carried him out, and turned him off
at some distance from the building.

Mr. Selby mentions another instance, in which a
most determined battle was fought between two red-
breasts, who were so engrossed with the combat, that
they allowed themselves to be twice taken up and
separated by a person witnessing it.   The occupation
of a shed seemed to be the object of dispute.

———

Feb. 12th, 1845. — White says of the redbreast and the wren, that they sing all the year, hard
frost excepted.*   Not even hard frost, however, will
always cause the redbreast to desist.   We heard
one singing this morning at 7 o'clock, when the
ground was covered with snow, and Fahrenheit's thermometer was only 13 degrees, being 19 degrees below
the freezing point.   Nor was it, as the poet describes
it at this season,

Content
With slender notes, and more than half suppress'd ;

but its voice was as strong, and its notes as varied
and melodious, as in the height of spring.

* *Nat. Hist. of Selborne,* Lett. II. to D. Barrington.

PETTYCHAPS.[*]

This species of warbler is far from infrequent about us, haunting gardens, black-thorn copses, and high hedges, though more plentiful some years than others. Its note is soft, possessing much variety, and particularly pleasing, somewhat resembling that of the black-cap, but not so loud and powerful: it is often continued for a very considerable time without intermission. The songster itself is not often seen, from its extreme shyness, and its habit of concealing itself in the thickest covert. In the early part of the season, I have occasionally observed it in sycamores, picking at the flower-buds of that tree, to which it seems partial. I never heard it before the 1st of May, nor after the 20th of July.

LESSER WHITETHROAT.[†]

The lesser white-throat is common in this neighbourhood. During the second week in May in 1822, I observed a pair of these birds busily employed in building their nest in a low yew-bush close to the house. The nest was completed about the 12th: the first egg was laid on the 15th, and by the 19th there were five deposited, when the bird ceased laying, and commenced sitting. I had many opportunities of observing both sexes: the hen engaged on the nest; the cock generally close by, and constantly re-

[*] *Curruca hortensis,* Bechst.   [†] *Curruca sylviella,* Flem.

peating the shrill shivering cry which so peculiarly distinguishes this species. When the hen was put off the nest, she did not fly straight away, but merely hopped into an adjoining bush, and remained flitting about my head, uttering all the while a harsh sharp chattering note; which on one occasion was made by the cock also, but it was quite distinct from the usual call above-mentioned. Once, when I approached very near the nest, for the purpose of looking into it, the hen bird alighted on a twig within a yard of where I was standing, and uttered its harsh chirp, shivering with its wings, as if in great alarm for the fate of her dwelling. A day or two after incubation had commenced, I found the nest gone; which circumstance stopped further observation. The nest and eggs of this species are figured, along with the bird itself, by Latham, in the first Supplement to his *Synopsis of Birds*.* The accompanying description is very correct, and exactly accords with the nest and eggs above noticed. It may be simply observed, that the wool, which he mentions as one of the materials, forms no great part of the nest, but is merely scattered here and there in small patches, so as to keep the whole together, which is of a very flimsy structure throughout.

WOOD-WARBLER.†

WHITE observes of this species, that it "haunts only the tops of trees in high beechen woods;" and Mr. Selby remarks also, that " it frequents natural

* P. 185, pl. 113. 　　　　　† *Sylvia sibilatrix*, Bechst.

woods, and plantations of old growth, and is seldom seen in hedges or brushwood, like the yellow wren." This accounts for its rarity in this neighbourhood, (common as it is in many other counties,) where we have no large woods, and where the beech is not indigenous. I have never heard above three or four individuals in the course of a long term of years. These have been mostly noticed in the beginning of May, generally for a few days only; which leads me to doubt whether the species ever breeds with us, and whether the above were not merely in their way to other localities in the kingdom more congenial to their habits.

WILLOW-WARBLER. *

May 21st, 1824.—WE found to-day the nest of a willow-warbler, in a tuft of grass on the ground, containing five eggs. The nest was large for the size of the bird, and nearly spherical, with a small opening on one side near the top. It was composed of dried grass, stalks, and coarse herbage, and profusely lined with feathers. The eggs were white, and speckled all over with light rust-colour; in some, the spots were chiefly confined to the larger end. They had been incubated some days.—I am told, in Norfolk the nest of this bird is called an *oven*, which it not unaptly represents in miniature, as regards shape.

This species of warbler will occasionally sing in its flight; Mr. Selby has observed the same of the wood-

* *Sylvia trochilus*, Lath.

warbler; last noticed : both these species, therefore, may be added to White's list of " Birds that Sing as they fly," * and which he states to be but " few."

CHIFF-CHAFF.†

THIS species, which is found in tolerable plenty in most parts of England, is of very uncertain appearance in this neighbourhood : in some seasons not a single individual is seen, whilst in others they are abundant. It is a restless and active bird, and seems much attached to spruce-firs and other tall trees, from the tops of which it issues its incessant but monotonous song, consisting of only two loud piercing notes ; which it continues throughout the summer, and sometimes even till late in September. It is most frequently heard here in the early spring, and then again in the autumn; which inclines me to think, as in the case of the wood-warbler above mentioned, that it does not generally breed with us, but merely stops here in its passage to and from other districts.

The chiff-chaff, in general, is only heard to utter the two loud notes above alluded to, which have gained for it its well-known name. In July, however, of last year (1845), I heard an individual varying its monotonous song, at intervals, with a kind of shrill whistle, repeated two or three times in succession. This whistling note was quite new to me, and such as I had never heard before from this

---

* *Nat. Hist. of Selborne*, Lett. II. to D. Barrington.
† *Sylvia hippolais*, Selb.

species; nor can I find it alluded to by any of our British ornithologists. I distinctly ascertained that it proceeded from the chiff-chaff, as no other bird was singing near at the time; and I could easily trace the individual that uttered it about the garden, as it flew from tree to tree.

WAGTAILS.*

I OBSERVE about here greater numbers of the common pied wagtail in the autumn than any other season of the year. This is probably accounted for by the circumstance of the birds of our own neighbourhood being joined at that time by those which arrive from the higher parts of the country. According to Selby, this species is a regular migrant in the North of England, retiring southward in October, and not reappearing till February or the beginning of March.

The grey wagtail † is most common with us in the month of January, and I know no instance of its being seen here in the summer.

TREE-PIPIT.‡

July 9th, 1827.—FOUND the nest of a tree-pipit placed on the ground beneath a tussock of *Aira cæspitosa*. It was formed of hay and grass, patched externally with moss and lined with fine grass. The

---

* *Motacilla yarrellii*, Gould. † *M. boarula*, Linn.
‡ *Anthus arboreus*, Bechst.

eggs were four in number, coloured greenish-brown with dusky streaks and blotches; their weight thirty-seven grains. They had been incubated some days.

From the above it would seem that this species is either a late breeder, or else sometimes has two broods in the season; most likely the latter. I have always noticed that it continues in song till some time after midsummer.

TITMICE.

THE notes of the titmice are rather puzzling to distinguish, owing to the variety possessed by some species, and the same note being occasionally used by more than one of them. The great titmouse* is the only one which has any pretensions to song. This species has several distinct sets of notes, all of them "joyous" and lively, which it repeats from the tops and higher branches of the larger trees. Besides these, it has two or three chattering notes, which it appears to possess in common with the blue titmouse† and others, and which are uttered principally whilst running up and down the smaller shrubs. I cannot discover that the blue titmouse itself has any one constant note peculiar to the species. The marsh titmouse‡ is sufficiently characterized by a note resembling the word *chip* or *chit*, repeated seven or eight times in pretty rapid succession, and which is often heard from the tops of tall trees in the early

* *Parus major*, Linn.

† *P. cæruleus*, Linn.   ‡ *P. palustris*, Linn.

part of the spring. The cole titmouse* also frequently utters from the tops of the spruce-firs a
compound note, which sounds like *titwee*, repeated
quickly about five or six times; the first syllable
being pronounced very short, but great stress being
laid upon the second. Both these species, however,
have other notes, not so well marked, some of which
they use in common with the great titmouse.†
The long-tailed titmouse‡ appears to be almost
mute, or to possess scarcely more than one rather
weak note, that acts as a sort of call to the young
birds, which, after leaving the nest, follow the old
ones throughout the season.§

April 17, 1830. — Found the nest of a long-
tailed titmouse, in a red cedar, about five feet from
the ground; of a very curious and singular form,
long and oval, about the size of a smallish melon,
with a small hole in the side, through which the
parent bird enters; constructed chiefly of mosses,
wool, and dry grass, having the outside beautifully
studded with lichens, and the inside thickly lined
with a profusion of down and soft feathers. The
nest contained ten eggs, about the size of a small

---

* *P. ater*, Linn.

† It is correctly noted by an observing naturalist, that the
great titmouse and cole titmouse "will often acquire or compound a note, become delighted with it, and repeat it incessantly
for an hour or so, then seem to forget or be weary of it, and we
hear it no more."—*Journ. of a Naturalist*, 3rd edit. p. 164.

‡ *P. caudatus*, Linn.

§ According to Selby, the notes of this species " in the spring
are more varied, and it can utter a pleasing, though low and short
song." This, however, I never heard.

bean, of a white colour, thinly sprinkled with rusty dots at the larger end.

SNOW BUNTING.*

Snow buntings occasionally occur with us during the winter months, and are often called *white larks.* I note this, from White's mentioning that white larks had been supposed to have been seen now and then about Selborne in the winter-time, and his suggesting that they might have belonged to this species.

COMMON BUNTING.†

The common bunting, which White speaks of as a rare bird at Selborne, is extremely abundant in the open corn-lands of Cambridgeshire, where it is called the *bunting lark.* In winter-time they collect in flocks, and frequently assemble in large quantities on the bare twigs of some tall thorn, where they sit by the hour together uttering their harsh monotonous notes, not deserving the name of a song. Varieties of this bird, more or less spotted and variegated with white, are not uncommon: they sometimes occur wholly white, and are then called by the common people *white larks,* like the species last mentioned.

---

* *Plectrophanes nivalis,* Meyer.   † *Emberiza miliaria,* Linn.

REED BUNTING.*

WHITE did not appear to be well acquainted with
the note of this species. In our fens, where it is
abundant, its simple and inharmonious song, if song
it can be called, may be constantly heard during the
breeding season; the cock bird uttering it at short
intervals from the top of some low bush, near to
where the hen is engaged in incubation. It consists
of two long notes, followed by three or four similar
but short ones,—these last being hurriedly pronounced.

SISKIN.†

Feb. 15, 1829. — SISKINS everywhere plentiful
just now. They seem most attached to alders, and
hang upon the boughs of those trees with their
heads downwards like titmice, whilst endeavouring to
get at the seeds of the last year's cones. The crops
and stomachs of some individuals which we examined were stuffed quite tight with this kind of
food exclusively, nor was there a single seed of any
other tree or vegetable to be found.

These birds are not unfrequent about Bottisham
most winters during the months of January and
February, but are more numerous this year than
usual.

* *Emberiza schœniclus,* Linn.   † *Carduelis spinus,* Steph.

RAVEN.*

On offering a dead bat to a tame raven, he
seized it with the greatest avidity, and devoured
every portion of it, not even rejecting the leathern
wing.

ROOK.†

Few persons who have lived much in the neighbourhood of an extensive rookery, as has been
the case with myself, have not something to say
in reference to the habits of these birds, which,
from living in large communities, strike the attention of the most careless observer. I am always
much amused in watching them of an evening as
they return to roost during the summer and autumnal months. After the breeding season is over, they
seem for a while to desert their nest-trees entirely,
and for a few weeks we see none there, except perhaps two or three stray individuals, by night or by
day. But about the end of June a few begin to
return home in the evening, and the numbers continually increase during July, August, and September, till at length the air is almost darkened by the
collected multitudes winding homewards at the decline of day.‡ The regularity, and I may say punctuality, with which they return is very surprising.

* *Corvus corax*, Linn.   † *C. frugilegus*, Linn.
‡ It is generally stated in ornithological works, that rooks do
not return to their nest-trees till the autumn. But what I have

At the lapse of a certain interval after sunset, they are sure to be heard giving notice of their approach by their united voices. The length or shortness of this interval depends upon the time of the year and the character of the weather. For the first fortnight after midsummer, when the twilight is strong till quite late in the evening, those that come home at all do not return till near an hour after sunset. The average time for their return during July I find to be forty-five minutes after sunset, if the weather be fine and clear; but, if much overcast, they generally shew themselves about ten minutes earlier.

In August as it gets more rapidly dusk, the interval is reduced to about thirty or thirty-five minutes after sunset; and in September it is still further reduced,—on an average to twenty-five minutes. From all this it would seem that they are probably guided by the light. Sometimes, when the weather is damp and gloomy, but nevertheless mild, I observe that they will return from feeding half an hour or an hour earlier than usual, but do not retire actually to the roost-trees till the accustomed time, employing that interval in flying about in the neighbourhood, or settling for a while on the trees that lie on the outskirts of their immediate dormitory. When they really betake themselves to rest, or for a few minutes

mentioned above in respect of their returning at night to roost, is certainly true in this neighbourhood, and the result of many successive years' observations. The numbers so returning are sometimes very considerable, even as early as the first and second weeks in July.

previous, they exercise their voices all together in a very peculiar way, different from at other times, producing a deep hollow sound that may be heard to a considerable distance; and it is a quarter of an hour or twenty minutes before they are hushed for the night.

Late in the autumn, if the afternoons are cold and frosty, and also often during winter, they may be observed sitting in a sullen way, in immense clusters, at the tops of the trees near adjoining the rookery, which are blackened with their numbers, for an hour or two before roosting: at such times they are quite silent, and hardly noticed by persons passing, till they all suddenly take wing, frightened at his approach. But, if the weather be mild and open, they keep flying about at the decline of day in immense multitudes, high in the air, with an incessant cawing, till the increasing darkness obliges them to retire for the night.

———

Rooks appear to be very restless at times in the dead of the night, and to wake much; for often at 10 and 11 p.m. (and this in winter as well as summer) I hear a squabbling and cawing on the nest-trees near my room-window. They are early risers too, and at all seasons leave their roosting-places half an hour or more before sunrise.

———

Jan. 16, 1830.—Rooks up this morning at half past seven exactly. *N.B.* Sky much overcast, with sleet and rain at intervals. In the afternoon, about two, seen in immense flocks closely compacted together, hovering over the lands adjoining Newmarket Heath, principally turnip-fields.

For a few weeks previous to the actual time of building, as early as the commencement of February, and sometimes earlier, if the weather be mild, the rooks remain upon their nest-trees during a great part of the day, as if in contemplation of the approaching season. Whilst so collected, they are very clamorous, and much confusion seems to prevail amongst them, as if they were trying to claim their old nests, or were disputing for those already in part made. At this period of the year, also, they are very shy, and easily alarmed; whereas in the advanced spring, when building, and the cares attendant upon bringing up the young brood, engross their attention, they seem comparatively to take little notice of persons passing under the trees.

Rooks, like many other birds, have a particular note which they utter when wishing to sound an alarm of danger to their companions. When the keeper at Bottisham Hall appears with his gun, approaching the trees on which they build, if a single individual see him, it immediately warns the rest of the community, who with one accord rise simultaneously from the trees with loud vociferations, apprehensive of danger, though as yet unperceived by many of them. Frequently, where the rookery is extensive, and the trees straggling from one another, the alarm is quickly propagated from party to party down a long avenue, and in less than two minutes the whole is in an uproar. This is most noticeable in February, when the rooks are much on the trees during the day, as just above observed, but the building season not actually commenced.

Rooks that are much shot at, like those at Bottisham, have the greatest possible dread of a gun. Often, when walking in the neighbourhood of the trees, I observe they suffer me to approach within a certain distance without taking fright, though they watch my movements narrowly; or are not much disturbed, as long as I continue walking on : but the moment I stop, and put even a walking-stick to my shoulder, as if about to fire at them, they are off directly ; and the distance at which this will operate, often considerably more than a hundred yards, is quite surprising.* It is also observable, that when they would be frightened by any one walking or stopping near the trees, they will often suffer a cart or a carriage to pass immediately beneath them,

* I am induced particularly to mention this circumstance, because it has been sometimes stated that they can distinguish a gun from a stick : and that a person with only this last, may feign as if about to fire at them, without putting them to flight.—See *De Selys-Longchamps, Faune Belge*, p. 69. Perhaps the habits of individuals may differ in different localities.

without taking the least notice, as if conscious that there was no danger to be apprehended from such a source.

———

The rooks seldom begin to build in good earnest before the first week in March. One year (1826) they were first observed busy in this way on the 4th of March; the young were first heard squeaking in the nests on the 7th of April, and the first fledged birds were noticed on the 29th of the same month. These dates give a period of exactly two lunar months allotted to the business of nidification, of which about five weeks are taken up in building and incubating, and the remaining three devoted to feeding and bringing up the young birds, till able to leave the nest.—In subsequent years, these dates have sometimes fallen a few days later, but the intervals between them respectively have remained nearly the same.

There are also always some pairs much behind the others, to the extent of a month or more. A few young rooks may be observed most years sitting at the edge of the nest, hardly yet fledged, as late as the beginning of June, long after the great bulk have flown and taken their departure.

———

In building their nests, rooks are not content with picking up any dead sticks that may fall in their way, but use principally fresh twigs, which they forcibly detach from the branches of the neighbouring trees. They may often be observed at such

H

times clinging to the twigs with their bills and claws, and pulling at thèm, at the same time flapping their wings in order to support themselves. The nests are lined with grass, and other vegetable matters of that kind. One year we noticed a number of nests having the leaves of pine-apples interwoven with the sticks; these had been obtained from a dung-hill in the garden near a hot-house, out of which the dead leaves of these plants had been ejected.

In large rookeries, during the breeding-season, a strong smell proceeds from these birds, very perceptible to persons walking under the trees on which the nests are built. This seems to arise partly from the number of individuals collected together, and partly from the quantity of dung dropped. Where this dung falls, nettles vegetate very luxuriously.

March 18*th*, 1843.—The nests of the rooks appear finished, yet few birds can be observed actually sitting. They keep stationary on the trees, perched in pairs on the very uppermost twigs, which bend beneath their weight. They spend much of their time in cawing, accompanied by a shaking of the wings; the hens receiving the attentions of the cock. Every now and then I see them pulling at the blossoms of the elms, on which trees the nests are mostly built.

I observe every year, that, from about the beginning of April to the end of the first week in May,
the rooks visit the pasture in front of my house
regularly every day. This is the period during
which the young are in the nest, and it is probably
for the purpose of obtaining food for them. They
chiefly come early in the morning, and are often all
gone by 8 o'clock A. M. With the help of a small
pocket eye-glass, I occasionally see some of them
walking about with their bills half open. Query, if
this is not owing to the number of grubs they suffer
to accumulate in their mouths before flying off with
their booty? They are constantly digging.

———

May 12*th*, 1828.—Rooks undoubtedly suck partridges' eggs occasionally, since the keeper at Bottisham, who found that havoc had been made in that
way by some vermin or other, lately set some traps
near the nests and caught several.

June 24*th*.—The men who are engaged in mowing
the hay and clover state that the rooks follow their
movements, and fall-to upon any nests which are exposed by the scythe, devouring the eggs without
mercy, and carrying others away in their bills,—not
sparing even those of the skylark. This singular
fact, which they say they never observed in former
seasons, seems to be a new circumstance in the
history of the rook, and confirms what was recorded
last month.

———

It has been sometimes thought that rooks have the power of discovering by scent, or some other faculty, the prevalence in any particular pasture of the grubs of the cockchaffer.* I should almost be inclined to doubt this, from the circumstance that, in the autumn of 1842, the field in front of my house was so attacked by these grubs that a large quantity of grass was completely destroyed; immense patches appearing as if scorched; and it was under consideration, whether it would not be necessary to pare and burn the whole pasture, and lay it down afresh; yet the rooks never found the grubs out. From my proximity to a large rookery, the inhabitants of which were daily under my observation, it was expected and confidently hoped each day that the rooks would soon discover and extirpate the enemy. To our surprise, not a single rook was seen to alight on any of the bare places; and, indeed, hardly a bird noticed in the field at all previous to Christmas, when the grubs had worked deeper from the approach of the cold weather, and were not easily obtained. The size of the pasture was a little less than three acres.

Dec. 30th, 1829.—Feeling some desire to know how rooks support themselves during severe frost, like that we have now experienced for this fortnight back, I caused one of these birds to be shot and

* See *Journ. of a Naturalist* (3rd edit.), p. 179. See also *Jesse's Gleanings in Natural History*, p. 62.

brought to me. It appears that they manage to subsist well, notwithstanding the cold weather; for, on opening the body, we were surprised to find it in most excellent condition, with the stomach, intestines, and other abdominal viscera, completely covered with layers of fat. The stomach itself was unusually distended, and projected externally below the extremity of the breast-bone like a large egg. The contents proved to be turnip and wheat, mixed with a few gravels. The turnip was in a semi-digested state, but easily identified by the smell: the grains of wheat were whole, and scarcely at all altered. The stomach was completely full of this mixed food, and stuffed as tight as a pin-cushion.

From the large size of the stomach in this individual, we were induced to measure it, and found it five inches in circumference, and two inches and-a-half in length. Whether this organ is always as large in the rook, and whether it is usual to find it so low down in the cavity of the abdomen as in the present instance, I am not aware; but it had a very remarkable appearance in this bird, and such as I never witnessed before in any other species. At the same time, it is possible that it may have been due to nothing else than to its being so unusually distended with food.

We made a few other anatomical memoranda with respect to this bird, as follows :—Pylorus eight lines below the cardiac opening;—œsophagus, with the lower part a little dilated, but, on the whole, of tolerably uniform character throughout its length ;— intestinal canal three feet long, furnished at the posterior extremity (about one inch from the vent) with two cæca, which lay parallel with the main gut, and were of unequal lengths, the longer one measuring eight lines, the other six :—the upper part of the intestinal canal, nearest the stomach, was of a much more red and florid appearance than the rest; this, however, may have been owing to some slight accidental inflammation.

————

Rooks will occasionally prey on fish. A relative of mine once saw a rook take a fish out of the piece of water in Kensington Gardens, and devour it on the bank. He told me he had seen jackdaws do the same in the Thames.*

————

Varieties of the rook occasionally occur here, as in other places. In the spring of 1823, we picked up two young birds alive under the nests, just fledged, in which each feather was tipped with dirty-white, giving the whole plumage a speckled appearance. One of them, also, had a single claw that was snow-white. These birds were found in different parts of the rookery, and did not appear to have belonged to the same brood.

* Mr. Blackwall has observed that the carrion crow also devours fish, " particularly eels, in pursuit of which it wades into the shallow water of rivers and brooks that flows over beds of stone and gravel ; seizing the object of its search with the bill and conveying it to land, where it is eaten at leisure."—*Ann. and Mag. of Nat. Hist.* vol. xv. p. 168.

The following year I had brought me a young rook, the plumage of which deviated from the usual character as follows: — Bill mostly of a pure white, having only a few black spots on the upper mandible; the forehead, including the feathers which fall forwards over the nostrils, mottled with black and white; chin wholly white; the first three quills on the right wing with a white line running longitudinally down each feather; the quills on the other wing not marked with this peculiarity;—the first four feathers on the greater wing-coverts, covering the primaries, wholly of a pure white on each wing; toes mottled with black and white; nails mostly pure white. All the rest of the plumage was of the usual colour.

In the spring of 1841, a completely white rook was shot near Cambridge. This, and the two spotted varieties first mentioned, are preserved in the Museum of the Cambridge Philosophical Society.

On the 12th of May, 1827, we picked up a dying rook at Bottisham, remarkable for a monstrosity in the bill; the upper mandible of which had grown to a præter-natural length, and passed beyond the lower to the distance of nearly an inch and a half. The individual was an old bird, but considerably under size, and had been apparently starved, being nothing but skin and bone. Doubtless the extraordinary prolongation of the upper mandible must have much interfered with the operation of grubbing in the earth, and prevented its obtaining more than a very scanty supply of food.*   Nevertheless the forehead and the base of the bill were as much denuded of feathers as in other adult individuals of this species.   This case of monstrosity seems analogous to that already instanced in the wild rabbit among quadrupeds.†

It is worthy of remark that this rook swarmed with lice to the most astonishing degree I ever witnessed in any animal.   Some parts of the plumage were quite alive and covered with them.‡   It is probable that all vermin increase greatly upon diseased animals, owing to their neglecting the ordinary means of ridding themselves of them ; and in some cases they may serve to hasten their death, and thus relieve them from their misery.

JACKDAW.§

JACKDAWS sometimes prove a great nuisance in this part of the country in spring, from their habit of building in chimneys.   They often bring such a number of

* See a representation of the cranium of a rook, having a similar peculiarity in the lower mandible, in one of the vignettes of *Yarrell's British Birds*, vol. ii. p. 90.

† See p. 78.

‡ These parasites consisted principally of the *Colpocephalum subæquale* of Nitzsch. See *Denny's Anoplura Britanniæ*, p. 213, pl. xviii. fig. 5.   I showed some of the specimens to this last gentleman, who has noticed the circumstance alluded to in the text, in his valuable and beautiful work.   With the above species of *Colpocephalum* there was also the *Docophorus atratus*, Nitzsch, but much less abundant. See *Denny*, pl. iv. fig. 8.

§ *Corvus monedula*, Linn.

sticks together as quite to choke up the flues. Whilst engaged in constructing their nests, they occasionally fall into the rooms below; and I have more than once been favoured with a visitor of this kind in my bed-room, in the early part of the morning. They do not seem much to mind smoke, as I have known them attempt to build in chimneys where there were fires kept pretty regularly from day to day. From the quantity of horse-dung which occasionally falls into the grates beneath where they are at work, I should suppose that they employ this material in some way,—perhaps as a lining for their nests.

A white variety of this bird was once shot at Bottisham.

———

April 27, 1828. — A farmer in this neighbourhood observes, that, for about a month at this period of the year, his corn-stacks are more resorted to and attacked by the jackdaws than at any other time, and that he is obliged to employ a boy to keep them off. This circumstance, one would suppose, must have some connection with the breeding-season.

———

Jackdaws are not only fond of carrion and offal, but will occasionally attack living prey. A friend of mine saw one well punished for attempting to take a young chicken just hatched. A cock and the old hen beat him so severely, that he made his escape with difficulty.

JAY. *

Nov. 20, 1826. — JAYS are not regular denizens
of this neighbourhood, but only visit it occasionally
at uncertain periods. A party of these birds, however, have frequented our plantations for several
weeks back; and one was brought me to-day, the
first that could be obtained, owing to their extreme
shyness. On examining the stomach of this bird,
we found it stuffed to the full with the remains
of both animal and vegetable substances. These
were in a state of intimate union, and had the appearance of a stiff mud; amongst which, however,
was discernible the tail of a mouse, (there being no
trace of any other part of the animal,) and a great
quantity of the seeds of some berry, staining paper of
a purplish-blue : there were also many gravels, some
of considerable dimensions. There were no traces of
either acorns or beech-mast, which we had supposed
to have been what had attracted these birds to
this neighbourhood.

GREAT SPOTTED WOODPECKER.†

THE great spotted woodpecker is of not very
unfrequent occurrence in this neighbourhood; but
it is remarkable that all the specimens that have
come to my knowledge have been killed either in
spring or autumn, and mostly in March. This

* *Garrulus glandarius*, Flem.   † *Picus major*, Linn.

looks as if it were migratory with us, at least from one part of the country to another, and that we saw it only *in transitu.* In March 1827, two male and female, were shot together in the park at Bottisham, but I never heard of the nest being taken about here; nor am I aware that they breed so early in the spring as to induce the belief that these had paired for that purpose.

The noise made by these birds is very peculiar, and one that I had often heard in the spring for several years in succession, before ascertaining the source whence it proceeded. It very much resembles the creaking of an old tree in the wind, though I never could observe the way in which it was produced.*

I have noticed that this species of woodpecker always swarms to a prodigious degree with a minute kind of *Acarus* (*Sarcoptes?* Latr.) mostly infesting the plumage about the head and neck. It is closely allied to the *Ac. avicularum* of De Geer†, but is specifically distinct.

COMMON CREEPER.‡

May 1, 1826.—To-day we found a creeper's nest beneath the loose and decayed bark of one of the limes in the avenue in Bottisham Park, a situation in which

---

* Some authors assert that it is produced by the bird putting the point of its bill into a crack of the limb of a large tree, and in that position making a quick tremulous motion with its head.

† *Mem.* tom. vii. p. 107, pl. vi. fig. 9.

‡ *Certhia familiaris,* Linn.

these birds are fond of building. This nest afforded
an instance of the long interval that occasionally
elapses with some birds between completing their
nest and commencing incubation;* as I had observed it apparently in a finished state, but without eggs, as long back as the 13th of last month.
To-day the old bird was sitting for the first time,
more than a fortnight afterwards. This nest consisted of twigs and small sticks piled rudely together, and somewhat compacted by the addition of
wool, mosses, and *jungermanniæ*, with a layer of
feathers on the top, and what appeared to be fine
shreds of wood probably torn from the inner surface
of the bark. Though tolerably firm and well matted
together towards the top, the nest was wholly supported by the bark; so that, if this had been hastily
broken away, it must have inevitably fallen. The
eggs were six in number, and much resembling those
of the willow-warbler, being of a pinkish white,
and marked at their larger end with numerous
rust-coloured spots. Their weight was seventeen
grains each.

The creeper has a pleasing, though somewhat
plaintive and monotonous song, which it utters from
the tops of trees early in the spring, and which might
be mistaken for that of some species of warbler,
were it not heard previous to the arrival of any of
our summer migrants.

* See p. 111.

NUTHATCH.*

Nuthatches are not uncommon in the neighbourhood of Bottisham, and may often be heard in the spring uttering their shrill whistling note. During a certain portion of the year these birds feed chiefly upon nuts, which they break with their bill, after having firmly fixed them in the crevices of the bark of trees. For this purpose they appear to resort frequently to the same spots, (as thrushes do to break snails,†) as I have observed some old trees in particular whose clefts are full of broken shells, whilst in others not one is to be seen.

———

March 6, 1843.—A nuthatch was shot by the keeper at Bottisham Hall to-day, while feeding, as he asserted, on a dead sheep. This seems in keeping with Bewick's observation respecting this species, that it is fond of picking bones; though I never before knew an instance of its shewing any such carnivorous propensities. The stomach, however, of this individual, which we examined, and which was very strong and muscular, did not appear to contain much of this kind of food. It was principally stuffed with gravels, and the remains of coleopterous insects, amongst which the elytra of ladybirds were very distinguishable.

* *Sitta europæa*, Linn. † See p. 127.

CUCKOO. *

THE note of the female cuckoo is so unlike that of the male, which is familiar to every one, that persons are sometimes with difficulty persuaded that it proceeds from that bird. It is a kind of chattering cry, consisting of a few notes uttered fast in succession, but remarkably clear and liquid.

KING-FISHER.†

KING-FISHERS are occasionally seen skimming over the surface of a stream which runs along one side of my garden. I am often warned of their approach by a shrill piercing note, which they utter during their rapid flight, and which may be heard to a great distance. I hear this note at all seasons of the year.

SWALLOW.‡

THE arrival of this species at Swaffham Bulbeck takes place on the 19th of April, according to a mean of twenty years' observations. The first broods are fledged in the middle of June, the second early in August, and towards the middle of the latter month these begin to collect into large flocks, which increase in numbers as the season ad-

---

* *Cuculus canorus*, Linn.   † *Alcedo ispida*, Linn.
‡ *Hirundo rustica*, Linn.

vances, and the time of departure draws near. This, with respect to the majority, takes place in the beginning of October; but stragglers (generally young birds of the second hatching) may be often noticed till the third or fourth week in that month.

I once observed a white variety of this bird at Ely, flying in company with others of the usual colour.

———

The swallow has a peculiar note, which it utters in the height of summer, just at break of day, when it begins to get light. This note differs from its ordinary song at other times, in being much less varied and lively, and is put forth in a peculiarly plaintive and very monotonous manner. It is sometimes repeated without intermission for nearly an hour together, and is always uttered from the nest or chimney top, while perched, and appears to be preliminary to taking the first morning flight. This note of the swallow is often the earliest of any bird heard in the morning, the skylark excepted : the thrush and the blackbird, however, strike up shortly afterwards, and occasionally take the precedency ; but all appear to be regulated by the degree of light. On very fine mornings I have heard the swallow singing at twenty minutes after two ; but when it has been wet, and the sky much clouded, I have not heard it till near three o'clock. After it has taken wing, (which is often not till an hour, or an hour and a half from the time of its first commencing the above note,) it then assumes its more usual song.

HOUSE MARTIN. *

HOUSE MARTINS with us appear about a week or
ten days after the swallow, but are seldom numerous
till the beginning of May. They, however, remain
with us later than that species, and are occasionally
seen through the first week in November, though
the greater part withdraw before that time. In the
autumn of 1842 they were seen by myself at Cambridge as late as the middle of that month.

————

Martins love to build under the projecting eaves of
my house; but so abundant are the sparrows in this
corn district, that more than half the number of
nests are seized, as soon as finished, by these last
birds, and adapted to their own use, by a plentiful lining of straws and feathers; the rightful pro-
prietors being forcibly expelled. The old line,

Sic vos non vobis nidificatis aves,

is here true in a more exact sense than the poet
contemplated.

It is principally in June and July, whilst engaged
in providing for the first broods, that the martins are
annoyed by these besiegers. Nests built in August
and September for the later broods generally escape
molestation, the sparrows by that time having mostly
reared their own young.

* *Hirundo urbica,* Linn.

————

Sept. 3rd, 1829.—House martins have a singular practice throughout the breeding-season, and more particularly towards the latter part of it, of flying up against the walls of buildings, just below the eaves, and daubing them with mud, apparently without any intention of constructing a nest. Perhaps they do not go twice to the same spot: at any rate, these patches of dirt are not applied with any regularity, but may be seen sticking to the brickwork, at intervals of two or more inches all along the front of the building. Just at the present time, my own house has a line of these mud patches carried round nearly three sides of it. I fancy I notice, that the birds are more inclined to this sort of proceeding in some states of weather than others. Occasionally, twenty or thirty martins will be busily engaged in this manner from morning till night, when perhaps, for several days before and after, not one is to be noticed. A damp, cloudy day, especially if also warm, seems to call them most to this employment, during which they appear actuated by some feeling or excitement which it is difficult to explain. It is surely something more than an instance of their " caprice in fixing on a nesting-place," (alluded to by White,*) which induces them to " begin many edifices, and leave them unfinished." In the present instance, I suspect they may be the first broods but lately fledged, whose instinct begins to operate and shew itself in this manner before it is wanted.

We see something analogous to this in the beha-

* *Nat. Hist. of Selborne,* Lett. XVI. to D. Barrington.

viour of caged birds, who seem to experience a pleasurable sensation in carrying about in their bills, at
the proper season for building, any little pieces of
thread, straw, or other materials, which happen to
be within their reach, even when they have no mate,
and no convenience for making a regular nest. This
is evidently an instinctive action occasioned by the
excitement of unsatisfied desire, increased perhaps
by high and stimulating food.

SAND MARTIN.*

I ONCE met with a person who assured me positively, he had in former years seen swallows taken
in a torpid state during winter from the holes and
crevices in the gravelly banks by the side of the road
near Quy Water. At this place, which is between
three and four miles from hence, I observe a small
colony of sand martins every year; and, if the above
fact be correct, which is very doubtful, it was probably this species of swallow which was discovered
there. Yet White mentions instances† in which
the holes of sand martins had been expressly searched
during the cold months, with a view to discovering
whether they used them as hybernacula, without
finding a single torpid bird.

The locality above mentioned is the only one I
know of in this neighbourhood, for many miles
round, frequented by this species of swallow; and
they are not numerous there. The soil being prin-

* *Hirundo riparia*, Linn.
† *Naturalists' Calendar* (by Aikin), p. 83.

cipally chalk or clay, with very little of gravelly or sandy deposits, is unfavourable to their habits.

SWIFT.*

SWIFTS, which delight much in large and extensive buildings, are not numerous in this village. Recollecting White's remark, that he observed the same number of pairs at Selborne invariably every year,† I have often counted the few that are to be seen flying about, in order to ascertain whether any thing similar is the case here. As far as my observation goes, the number does vary, and seems to have been rather increasing of late years. Previously to 1826, I never noticed more than two pairs; but since then I have observed generally four, and during the one or two last seasons as many as five. These all build in the crannies of the church steeple, with the exception of a single pair that take up their abode from year to year under the eaves of a dovecote in an adjoining orchard. They are always late in coming to us in the spring, being seldom seen before the middle of May, and are generally gone again by the end of July; thus making but a short visit of scarcely more than nine or ten weeks.

———

Swifts remain longer at Ely than any other place where I have had an opportunity of observing them. The bulk of them hardly ever retire there till quite

---

* *Cypselus apus,* Flem.
† *Nat. Hist. of Selborne,* Lett. XXXIX. to D. Barrington.

the end of August; and a few individuals may often
be observed through the first week in September.
From what cause they are induced to protract their
stay at that place so much beyond its usual limit, I
am unable to say; but the fact itself I regularly
noticed during a period of several years that I was in
the habit of residing there for the summer months.
Possibly they may, in some measure, be influenced
by the cathedral and other old buildings adjacent, in
the holes and crannies of which these birds meet
with a retreat peculiarly congenial to their habits, as
appears by the immense numbers that annually resort
thither in the early part of the season.

RING-DOVE.*

RING-DOVES are very abundant in this neighbourhood, though less so than formerly, and do much
mischief some seasons by devouring peas, beans, and
other leguminous plants. Their cooing notes are
heard incessantly from February to October. After
that time they begin to collect into enormous flocks,
which disperse themselves over the country during
the day-time to feed, but return regularly home in
the evening to roost in their native woods and plantations. Some of these flocks do not separate till
very late in the spring, though the greater part pair
off for the purpose of breeding by the beginning of
March. In the autumn, they appear to subsist
chiefly on acorns and beech-mast.

* *Columba palumbus*, Linn.

The coo of the ring-dove consists of five or six distinct notes, the whole being repeated generally three times, occasionally four or more ; but, what is curious, it almost invariably stops with the *first* note in the series.

Feb. 6*th*, 1830.—During the severe weather which has prevailed of late, the field in front of my house, which has the remains of a crop of cole-seed upon it, has been visited daily by large flocks of ring-doves, which subsist upon the leaves, and seemingly draw their support entirely from this plant.

A singular circumstance occurred to my nephew one day last July (1845). As he was walking in the fir-plantations at Bottisham Hall, a ring-dove fell suddenly to the ground a short distance from him, as if wounded. On going up to it and securing it, he found it swarming with individuals of some species of fly, which, by his description, I have no doubt was a species of *ornithomyia*, and which appear to have collected upon the bird in such numbers as quite to overpower it. It was a young bird of the year, but fully fledged, and had not fallen from any nest; neither did it appear to have sustained any injury in other ways.

STOCK-DOVE.*

WHITE mentions the stock-dove as being seen at Selborne during the winter only, appearing in large

* *Columba œnas*, Linn.

flocks about the end of November, and departing in February. With us, however, they certainly remain the whole year, as I have noticed them at all seasons and repeatedly found their nests. They are by no means uncommon in the plantations at Bottisham Hall, though less plentiful than the ring-dove. Their habits are very similar to those of this last species, with which they frequently associate in hard weather. Like them, they breed very early in the spring. The nest, which is flat and shallow, consists merely of a few sticks put loosely together in the hollow of some old tree. The eggs are two in number, white like those of the ring-dove, but somewhat smaller and rather more rounded. The stock-dove does not coo like the ring-dove, but utters only a hollow rumbling note during the breeding-season, which may be heard to a considerable distance. Both species are indiscriminately called *wood-pigeons* by the country-people.

March 21st, 1822.—A stock-dove was shot at Bottisham this evening as it was returning from feed. On handling it, a distinct rattling noise was heard in the crop; from which, when opened, we took out nearly a teacup-full of field-pease, above fourscore in number. They were unchanged in any way, excepting that some appeared to have sprouted since they had been swallowed.* After they were taken out, they

* A similar fact is recorded with respect to acorns taken from the crop of a wood-pigeon ; many of which, in the instance mentioned, are said to have " been so forced by the warmth and mois-

retained for a considerable time a strong vinous smell.

———

May 4th, 1827.—A stock-dove's nest was found to-day in a hollow tree in Bottisham Park, and the old bird shot as she flew off. This nest had been constructed in an old pollard elm that was hollow for about twelve or eighteen inches down from the top; but the top itself was covered over, and the sides also enclosed, so as to afford the bird no entrance except through one of three or four holes, about the size of itself, opening outwards in different places. The surface of the decayed wood, or floor of this cavity, which somewhat resembled in shape a small oven, was strewed over with earth, bits of stick, a few dead leaves, and other rubbish; but appeared to have received very few additional materials brought by the bird, with the view of forming much of a nest properly so called. The eggs were two in number, and had been incubated some days : they weighed four drachms and twelve grains each.

———

Sept. 28th, 1827.—Both ring-doves and stock-doves are subject to a peculiar kind of disease, that shews itself in fleshy tumours and fungous concretions somewhat resembling warts : these affect more especially the feet, bill, sides of the head, and contour of the eyes.—To-day we picked up a young stock-dove in a dying state, which had those parts completely

ture of the stomach as to have thrown out roots." See *The Zoologist*, p. 649.

covered with such excrescences, many of a large size.
They were more full out than in any specimen I had
ever seen previously.

COMMON TURTLE-DOVE.*

I<small>T</small> is singular that turtle-doves should never have
been seen or heard by me in this neighbourhood till
1823.   That year, and the year following, we noticed
a few individuals : since then the numbers have been
increasing each season; and now their very peculiar
notes, more resembling the whizzing noise of a spin-
ning-wheel than anything else, may be heard everywhere during the spring and summer from the middle
of May till after the solstice.   They appear to be
partial to fir-plantations.   Whether the old birds
continue with us after the breeding-season is over,
I am not certain; but I have occasionally known the
young shot in the month of September.

On the 18th of June, 1824, we found the nest of a
turtle-dove in one of the plantations at Bottisham
Hall.   It was built of sticks, and shallow; like that
of the ring-dove, only smaller.   There were two
eggs, of a pure white, and somewhat elliptical; also
resembling those of the ring-dove, except in size.
They measured about an inch and a quarter in length :
they did not appear to have been incubated.

COLLARED OR AFRICAN TURTLE.†

T<small>HE</small> unfledged young of all the pigeon tribe are

---

* *Columba turtur*, Linn.   † *C. risoria*, Auct.

fed, as is well known, from the macerated contents of
the crop of the parent birds, mixed with a curdy
secretion of the crop itself. I am often much amused
in watching the way in which this is effected in the
instance of the collared or African turtle, of which I
have several individuals in confinement. The old
bird opening its beak to the full extent, the young
plunges its own almost, as it were, down the throat
of its parent, whose efforts to regurgitate the required
food into the mouth of its offspring are distinctly
visible. But what particularly takes my attention is
the persevering and often fruitless endeavour made
by the young to induce the parent to open its mouth
for this purpose. This is especially the case when
the young are now nearly fully fledged, and partially
able to feed themselves; and when, perhaps, the
usual secretions of the parent's crop are beginning to
fail. Under such circumstances, they will often
chase the old birds round the cage, and again and
again present themselves before their face, as often
as they turn away from their solicitations: at the
same time they keep up a continual flapping with
their wings, utter a plaintive whining note, and
peck at the sides of their parents' bill, trying every
stratagem to make them yield to their entreaties.—
The old birds, however, as if conscious that there
was no supply, or that it was no longer needed, obsti-
nately refuse to pay any regard to the demands made
upon them; or they are not prevailed upon till after
a long time, and till wearied, as it were, with the
perpetual teazings of their offspring. The difficulty
experienced by this last in effecting its object is

I

greater as it advances to the age at which it is capable of taking care of itself. Probably the secretion in the parent's crop is dependent upon a certain degree of excitement caused by maternal affection; and, after a time, when this excitement wears off, by reason of the increasing age of the young bird, it is with difficulty elaborated. At length it ceases altogether; yet the habit of the young coming to its parent to be fed is kept up for a while, in like manner as we see nearly full-grown kittens and puppies still occasionally pulling at their mother's teats after they are dry. The scene above described may, at any time, be witnessed by throwing down a little hemp-seed into the cage where the parent and young birds are, when, as soon as ever the former begin to feed, the latter will be immediately at them importuning for a share.

DOMESTIC COCK.

" The cock's shrill clarion" does not always wait for the break of dawn, but is occasionally sounded during the dead of the night, and this even in winter; when, if the air be still and frosty, the distance to which it may be heard is quite surprising. I have at such times distinctly heard two cocks calling to one another from two different homesteads, situate a mile and a half or more apart.* The notion of cocks

* Mr. Blackwall has observed, that the hooting of the tawny owl may be heard to the distance of a mile, or even two miles, under very favourable circumstances. See *Ann. and Mag. of Nat. Hist.* vol. xv. p. 167.

crowing often during the night before Christmas, and its supposed effect of driving off evil spirits, is familiar to us from the well-known lines in Hamlet :

> Some say, that ever 'gainst that season comes
> Wherein our Saviour's birth is celebrated,
> This bird of dawning singeth all night long :
> And then, they say, no spirit dares stir abroad ;
> The nights are wholesome ; then no planets strike,
> No fairy takes, nor witch hath power to charm,
> So hallow'd and so gracious is the time.

I once was staying at a friend's house in the month of January, where the cock crew every evening regularly at nine o'clock; keeping it up for about ten minutes, and then desisting.

March, 1841.—A friend, residing in the next village to this, communicates to me the following circumstance in proof of the courage and ferocity of the common hen under the attacks of an enemy, whilst engaged in incubation.

A fowl, which had commenced sitting in the henhouse upon thirteen eggs, was observed each successive morning to have lost one or more of them during the night, till the number was reduced to nine. At length, one morning, a rat was found lying dead on the ground near the nest, with its skull fractured, whilst the hen bore marks of having sustained a severe conflict: her breast was torn and bloody, and her feathers much ruffled. The rat was a very large one ; and there could be no doubt of its having been killed by the hen, on renewing its attempt to get at more of the eggs.*

PHEASANT.†

May 24, 1827.—A HEN PHEASANT was picked up this morning, near its nest, dead, but still warm. There was no appearance of any external injury. On opening the crop, we found it stuffed quite full with the blossoms and roots of the common butter-cup (*Ranunculus bulbosus*), which had been swallow-ed whole, and did not appear to be the least changed. At first we thought this circumstance had something to do with the death of the individual; but upon inquiry learnt that these birds were in the habit of feeding upon the roots of this plant, strong and acrid as it is, during the spring and summer months. ‡

The cock pheasant sometimes exhibits marks of great daring and fierceness, even attacking man. I was once staying with a friend, who had a bird of this character in the plantations near his house,

* A somewhat similar case of a "combat between a hen and a rat" is recorded in *Loudon's Mag. of Nat. Hist.* vol. ix. p. 105.

† *Phasianus colchicus*, Linn.

‡ I find Mr. Selby also mentions, in his *Illustrations of British Ornithology*, that he has observed that the root of the *bulbous crowfoot* (Ranunculus bulbosus) is particularly sought after by the pheasant, and forms a great portion of its food during the months of May and June.

which was accustomed to make frequent sallies
upon persons passing near the place of its resort.
I saw it myself fly boldly at the proprietor of the
grounds, who purposely approached the spot, in
order that I might witness the extent of its courage and ferocity;—it commenced pecking his legs,
and striking with its wings, pursuing him for a considerable distance down one of the walks. He said
that he generally carried a stick to beat it off, whenever he went that way. Some wood-cutters, who
were at work close by, were in the habit of pro-
tecting their legs with strong leather gaiters from
the attacks of this bird, which was constantly in-
terrupting and annoying them in this manner.

PARTRIDGE.*

June 21, 1824. — A BROOD of young partridges
went off this morning. This species appears in
general to hatch about a week or ten days later
than the pheasant.

A covey of these birds were bred in this neighbourhood a few seasons back, the greater part of
which were perfectly white.

Sept. 30, 1829.—A FRIEND, who was lately sporting
in the sandy parts of Suffolk, observes that the partridges about there are so infested this season with
lice, that numbers have died in consequence. He

* *Perdix cinerea*, Lath.

suggests it may be owing in some measure to the wet summer, which has kept the soil in such a wet state as to prevent their dusting themselves so effectually as usual.

———

The courageous and almost fearless conduct of birds during the breeding season, even in the case of species at other times remarkable for their timidity, is very striking, though it has often attracted the attention of the natural observer. I have, a little way back, mentioned an instance of such daring courage in the domestic hen. Mr. Selby sends me another, in the case of the partridge, which occurred in his neighbourhood, and which was observed by a person on whose veracity he can rely, and who indeed produced evidence quite conclusive of the nature of the scene of which he had been an eyewitness.

Happening to be walking in a grass-field one day in the beginning of July, his attention was arrested by cries and screams, which he soon recognized as those of a partridge in distress, when alarmed for the safety of her young brood. On looking to the quarter whence the cries proceeded, he perceived two partridges engaged in a severe conflict with a carrion-crow, which, no doubt, had made an attack upon the newly hatched young, and attempted to carry some of them off. This the parents resented with such determination and vigour, as not only to prevent the crow from executing his intentions, but to compel him to stand

on the defensive.   They continued persecuting their
enemy; and, not content with driving him fairly off
the field, carried on the assault, till the crow was so
fatigued with defending himself from their attacks,
and became so disabled by the blows he had received, as to be no longer equal to taking wing.
In this state he was eventually laid hold of by the
person witnessing the battle, who had quietly advanced towards the scene of action, unheeded by
the combatants, whose attention was wholly engrossed with the momentous struggle going on between them.

RED-LEGGED PARTRIDGE. *

ONE of these birds was shot near Anglesea Abbey, on the 27th of September 1821, previously to
which there was no instance to my knowledge of
its having occurred in this neighbourhood; but since
then many other individuals have been met with, and
the species appears to be getting more and more
plentiful each year.   In 1839, during the last week
in June, a nest was found containing numerous eggs,
which were taken, and hatched under a hen in Bottisham Park.   This kind of partridge is tolerably plentiful in some parts of Norfolk and Suffolk, whence
probably our Cambridgeshire birds have originally
come.   We have occasionally dressed them for the
table, but found the flesh decidedly inferior to that
of the more common sort.

* *Perdix rubra,* Temm.

QUAIL. *

QUAILS are less plentiful with us than formerly, since the inclosure of so many of our open lands. They sometimes stay the winter, as in that of 1825-6, when they were shot about Bottisham, at intervals, throughout the whole of that season.   The whistling call-note of the male, heard occasionally in our corn-fields during June, is very peculiar, and often puzzles persons who are not familiar with it, to know from what it proceeds;—it frequently appears to come from some spot close to one's feet; yet it is almost impossible to get sight of the bird that utters it, from its sculking habits and its never taking wing at such times, but only retiring a little from the approach of the observer. This note I think, as far as my observation goes, is mostly heard of an evening; but I have sometimes heard it in given spots in certain fields without intermission from morning to night, the birds neither tiring with their monotonous exercise, nor seemingly wandering from a favourite station.

BUSTARD.†

Dec. 26, 1827.—A FRIEND assures me that he saw a bustard to-day in the open lands, between the village of Swaffham-Prior and Newmarket Heath, and that it approached sufficiently near for him to have

---

* *Coturnix vulgaris*, Flem.   † *Otis tarda*, Linn.

killed it, if he had had his gun with him. It is now many years since one of these noble birds was last seen in this neighbourhood. Persons are living, however, who remember them to have been by no means infrequent.

GREAT PLOVER. *

Sept. 29, 1830. — MY brother, while shooting to-day near Alington Hill, observed his dog make a point, and, on going up, found a specimen of the great plover, or stone curlew, which he secured alive. It proved to be a young bird of the year, which had in all probability been bred in the neighbourhood. After it was killed we opened it, and found the stomach to contain gravels mixed with the half-digested remains of coleopterous insects, amongst which the legs and elytra of some of the larger *carabidæ* were very conspicuous. These birds are not very uncommon about here some seasons; and in spring I occasionally hear their shrill startling cry, as they pass over the village late in the evening. The earliest period I have known them to occur is the 3rd of April.

DOTTEREL.†

May 7, 1829. — A NEIGHBOUR of ours shot some dotterel this morning on the open lands about Great Wilbraham, which were sent to me for exa-

* *Œdicnemus crepitans,* Temm.
† *Charadrius morinellus,* Linn.

I 5

mination. These birds are seen with us in the
spring and autumn only, for a short time, on their
passage to and from the North, where they breed.
The largest flocks generally occur in September.
Those killed in the present instance consisted of
one male and three females. The former, though not
an old individual, was in his full spring plumage,
and made a lively appearance. In the females the
colours were similar, but less bright and fixed.

On opening these birds, we found them loaded
with fat, which lay spread in broad bands over the
surface of the breast and abdomen: here and there
was noticed a peculiar oily secretion, intermixed
with the fat, and collected in large drops, which
floated on water. The liver was large, concealing
the stomach, and consisted of two lobes, the right
one nearly double the size of the left. The lower
part of the œsophagus was considerably dilated.
The stomach was a true gizzard, with the coats
very strong and muscular, varying in thickness
from two to four lines; it was filled with the elytra and other remains of coleopterous insects, especially the smaller *carabidæ*, intermixed with a few
gravels. The length of the intestinal canal was
sixteen inches and a half; its diameter nearly uniform throughout: at its further extremity were two
cæca, two inches in length, and taking their origin
about that distance from the vent.

In the female birds the ova were found to be
scarcely at all developed. It is probable, therefore,
that this species is a late breeder; and that, though
found on our open lands to the middle or even the

end of this month, it continues its journey to a much higher latitude before it rests for the business of nidification.*

We dressed two of these birds for the table, and found them exceedingly well-flavoured, and much resembling the snipe.

RINGED PLOVER.†

GREAT quantities of these birds appeared in our fens in the months of June and July 1824; but they have not occurred since, to my knowledge. The spring of that year was remarkably wet, so that a considerable portion of our low lands were completely flooded. There were at the same time several other species of water-birds observed there in plenty, which had not been seen for many years, and which were supposed to have quite deserted our neighbourhood. Amongst them were ruffs and reeves, godwits, redshanks, dunlins, and black terns. The appearance of our fenny districts that season, and the variety of birds that were attracted thither, gave us some idea of the state in which the fens habitually were in former times before drainage and culture had done so much to alter their character.

* Since the date of the above observations, Mr. Heysham has found this species breeding on the mountains of Cumberland. According to his account, they vary much in the time of laying, but the greater part are said seldom to commence before the first or second week in June. See *Yarrell's Birds*, vol. ii. p. 394.

† *Charadrius hiaticula*, Linn.

OYSTER-CATCHER.*

March 28, 1827.—Dr. Thackeray, the Provost of King's College, informs me that he obtained a few days since, in the Cambridge market, a specimen of the oyster-catcher, which had eggs in the ovarium in a state of great forwardness. This looks as if this species were an early breeder.

CURLEW. †

An ornithological friend informs me, that, in the spring of 1827, he found a curlew's nest in an old chalk-pit at Gogmagog Hills, and that the eggs are now in his possession. This is not a species of bird that I should ever have suspected of breeding in these parts.

GREEN SANDPIPER. ‡

Dec. 28, 1825.—The keeper at Bottisham Hall went up early this morning to Whiteland Springs, about a mile from this village, to look for ducks. On his return, he told us he had seen in the shallow streams about there what he called a *stone-runner*, but was not able to get a shot at it.—Not knowing what bird he alluded to under this name, we sent him back in the hopes of his finding it again, which he did, at the same time succeeding in killing it. It

* *Hæmatopus ostralegus*, Linn.   † *Numenius arquata*, Lath.
‡ *Totanus ochropus*, Temm.

proved to be a green sandpiper, a species which I had known to occur about here in a few previous instances in spring and autumn; but never at any other season. From the case of this individual, however, it would seem as if they remained the winter with us: whether they continue also during the summer, and breed anywhere in the neighbour-hood, I am not aware. On dissecting this bird, we were surprised at finding it so fat, especially as the stomach was quite empty, without a vestige of any-thing in it that could lead to a conjecture as to the nature of its food: the coats of the stomach were very strong and muscular.

COMMON SNIPE.*

June 15th, 1837.—A MAN brought me to-day an old snipe, with two young ones in their nestling fea-thers, which he obtained in Burwell Fen yesterday. —These birds are now known to breed in our low grounds in considerable numbers; but I never could hear of any instance of the nest of the jack snipe being found there.

DUNLIN.†

THESE birds occasionally visit our fens during the summer months, and I am inclined to think sometimes breed there. In the beginning of July, 1824, they were very abundant. Several which were then killed,

* *Scolopax gallinago*, Linn.
† *Tringa variabilis*, Temm.

and came under my observation, were in their summer or breeding plumage, as described by authors. The black, however, on the under parts was very variable in different specimens; some being only faintly spotted with this colour, others having the whole of the belly and abdomen thickly blotched over with large irregular patches of the same; but in no case without some mixture of white. Query, whether the season of incubation was not just over with these individuals, which had the two colours thus mixed, rather than not yet commenced?

WATER-RAIL.*

WATER-RAILS are common, as might be expected, in our fens and low grounds, and certainly are not migratory with us; at least they do not all migrate, as I have known them killed at all seasons of the year. Even the severest weather does not oblige them to quit the neighbourhood, as I have a note of one being shot at Bottisham on the 26th of January, 1827, during a very sharp frost, accompanied by deep snow, which had prevailed for more than a week previous. The evening before this bird was killed, the thermometer had descended to 12° Fahr., which is a degree of cold seldom exceeded in these parts.

I never found the nest of this bird myself, which, indeed, few naturalists have seen; but a man once brought me the eggs, which he had found in Bur-

* *Rallus aquaticus*, Linn.

well Fen : this was on the 8th of July, 1831.—These eggs were not very unlike those of the landrail, but of a whiter hue, with the spots smaller and much less numerous.

The body of the water-rail is strongly compressed, and admirably adapted for enabling the bird to insinuate itself between the reeds and high stems of sedge, which abound in the localities which it mostly haunts. The legs are long, and placed far behind, with a small space bare above the knee ; the toes likewise are long, and divided to their very origin. The wings are remarkably short, and but ill calculated for flight. Annexed to the bastard winglet is a small but sharp spur, two lines in length.—Query, if this is equally developed in both sexes ?

LANDRAIL. *

SOME seasons landrails are not uncommon in the grass meadows adjoining this village, and may often be heard in the evening uttering their peculiar harsh cry ; in others, not one is to be found. One thing always very much strikes me whilst listening to this bird, and that is the wonderful rapidity with which it seems to change its place in the field, judging from the note, which, from being one minute close at hand, in a few seconds will be so distant as to be almost out of hearing. Landrails are noted runners, and thread their way through the long grass with great swiftness ; but whether they really have the

* *Crex pratensis*, Bechst.

power of transporting themselves from one spot to another as quickly as it appears, or whether there may not be some deception in this matter, is doubtful.*   The note of this bird is sometimes heard during the day, but more frequently in the evening: during the first half of July one year, I noticed that it commenced pretty regularly about half-past nine, and was heard on to near midnight.

June 18th, 1827.—To-day, whilst mowing the hay in the field in front of the vicarage, we found the nest of a landrail containing seven eggs: these were of a very light brown, spotted and stained with rust-red, and not very unlike those of the moorhen; of an oval shape, and about an inch and a half long ; their weight 3 drachms 27 grains each.   They were fresh laid, and not as yet incubated.   The nest was little else than a hollow in the ground scantily lined with weeds and dead grass.   The note of the old bird had been heard for many days back in the neighbourhood of the spot where the nest was found.

SPOTTED CRAKE.†

I HAVE known these birds killed in Bottisham Fen as early as the 26th of March, and occasionally during the summer and autumn on to October, but

---

* Mr. Selby thinks that the bird varies its note in such a manner as to cause it to seem to a listener to come from different distances, producing thus an effect similar to ventriloquism.

† *Crex porzana*, Bechst.

not during the winter. They appear to be summer migrants. The body of this species of crake, like that of the landrail, is so strongly compressed, that when dead, and laid on its back on a table, it will not remain in that position, but rolls over to one side.

LITTLE CRAKE.*

A FEMALE specimen of this rare crake was caught alive in the fields at the back of Barnwell, near Cambridge, towards the end of March 1827.† Dr. Thackeray, the Provost of King's College, into whose collection it passed, informed me that, when opened, the eggs were found to be in a forward state. This indicates the species to be an early breeder, and looks as if the present individual might have bred in the neighbourhood, had it been suffered to remain.

MOORHEN.‡

MOORHENS occasionally build in trees. In one instance that occurred in the park at Bottisham, I found the nest constructed amongst the ivy encircling a large elm which hung over the water's edge, at the height of at least ten feet from the ground.

These birds not unfrequently appear on the lawn

---

* *Crex pusilla*, Selb.

† As already recorded by Mr. Yarrell, in his *British Birds*, vol. iii. p. 16.

‡ *Gallinula chloropus*, Lath.

in front of my house, picking up worms and insects within a few feet of the windows. When alarmed, they retire to a stream which runs along one side of the garden.

WILD GOOSE.

THE flocks of wild geese which visit our fens and corn lands in the winter season, and which consist principally of the bean* and white-fronted † kinds, are much less numerous than formerly, when scarce a day passed without more or fewer flying over the village:—now it is only occasionally that we see or hear them at all. This is attributed by some persons to the circumstance of rye crops being much diminished, in consequence of the improvement which has taken place in the cultivation of the land, enabling it to bear wheat in many places where only rye could be grown before. It has been observed, that these birds are always much attracted by young rye; the tender blades of which they devour with avidity, and which, being sown earlier than wheat, is often in a state of great forwardness by the end of October or beginning of November. One individual assures me that the enormous flocks which he used to see about thirty years back, at this period of the year, on the rye-lands in Great Wilbraham parish almost exceed belief. Very little rye is now grown there, or anywhere else in this neighbourhood, and no such flocks are seen at the present day.

* *Anser segetum,* Steph.   † *A. albifrons,* Steph.

TAME SWAN.*

Jan. 2nd, 1826.—M<small>Y</small> brother this morning seeing a tame swan, which frequents the piece of water
in the park at Bottisham, with something singular in
its mouth, approached it nearer in order to observe
what it was, when, to his surprise, he found it to be
a small roach, which the swan was dashing against
the surface of the water, and tearing in pieces with
its bill. On being disturbed, the swan let it drop,
and left it.—Whether this fish had been caught by
the swan in the first instance, or found dead and
floating upon the water, it would be interesting to
know. It was, however, evidently proceeding to
make a meal of it; and this is the first instance that
ever came to my knowledge of swans preying upon
fish at all.

DOMESTIC DUCK.†

Nov. 1829.—I <small>WAS</small> lately shown at Cambridge
three varieties, or rather monstrosities, of the common duck, which had been bred in the neighbourhood in the spring of the present year. These birds
deviated from the usual conformation of the duck
tribe in respect of their bill and feet. The former
had the lower mandible twice the length of the upper,
which last was deformed, and somewhat like that of
the common fowl: the tongue was long, and hanging

* *Cygnus olor*, Steph.
† *Anas boschas*, Linn.; *var.* domesticus.

out of the mouth. The legs appeared rather longer than usual, and not set so far back as is generally the case in ducks; and, what was principally remarkable, the toes were divided to their origin and without webs. These birds were in the Botanic Garden when I saw them, where there is a pond; but they took to the water with some reluctance, and but seldom resorted to it, as I was told, excepting when approached too nearly. They seemed to swim with some difficulty, and were compelled to keep their feet in constant motion in order to support themselves above the surface; sinking deeper than ducks usually do, with the breast nearly all immersed. One of the three individuals, which was a drake, and more active and healthy than the other two, occasionally dived half the length of the pond. —They all endeavoured to devour the *lemnæ* and other floating weeds on the surface of the water; but, from the monstrous formation of the bill, they did not readily succeed in securing their food.

The person to whom these birds had originally belonged, fancied they were a cross between the domestic cock and the common duck; and stated the circumstance of a cock and duck having been for some time alone in the farm-yard in which they had been reared, and which had led to an intimacy between them. There was nothing, however, in support of this idea to set against the extreme improbability of any fertile union taking place between two species of birds belonging to such totally different families. The absence of webs between the toes, which was the main point in which these individuals resembled the gallinaceous tribe, was clearly nothing more than a case of accidental monstrosity.*

WILD DUCKS.

DURING a short stay in the neighbourhood of Crowland in Lincolnshire, in August 1843, I was taken by a friend to see a duck decoy near that place. As decoys are now less plentiful than formerly, as well as becoming yearly less and less profitable, from the extensive drainage and cultivation of marsh lands, I was pleased with the opportunity of seeing one. This decoy covered twenty acres, but had not more than three acres of water in the middle. At the time of our visit, there were only a few wild ducks in it, supposed to be young birds that had been bred in the fens in the neighbourhood. The regular season, we were informed, begins in November, and ends in February or March. The best weather for taking the ducks is moderate frost and snow : in very mild wet winters they are too shy. The occupier of this decoy told us that, in his father's time, he had known as many as seven hundred dozen birds taken there in one season. This is, perhaps, as large a quantity from one decoy, during a single season, as any on record. Pennant mentions thirty-one thousand two hundred

* The above is the substance of some notes I made at the time of seeing these ducks. A short communication respecting them was afterwards published by some other person in *Loudon's Mag. of Nat. Hist.* vol. vii. p. 516.

ducks as having been sent up to London in one
winter from ten decoys together;* but this number,
when divided amongst the ten, is not more than
two hundred and sixty dozen each. Of late years,
the numbers taken in the decoy above spoken of
have greatly decreased, from causes already mentioned.

RED-BREASTED MERGANSER.†

Aug. 24th, 1840.—A MAN brought me to-day
alive an adult female of the red-breasted merganser,
which he had taken in Swaffham Fen.—This is a
species which is occasionally met with on our rivers
and streams during winter, but which I should never
have looked for at this season of the year.—These
birds are generally supposed to retire northward in
summer; and are known to breed in various parts of
Scotland and the Scotch islands. Mr. Selby found
them plentiful in Sutherlandshire in the month of
June, having then just commenced incubation. Is
the present individual likely to have come so far
south since rearing its young in those high latitudes,
or has it been breeding anywhere in this neighbourhood?—This last, if ascertained to be a fact, would
be an interesting discovery.

* *Brit. Zool.* (edit. 1812), vol. ii. p. 262.
† *Mergus serrator,* Linn.

SMEW.*

Jan. 16*th*, 1828.—A SMALL flock of smews appeared in Swaffham Fen a day or two back; one of which, a fine adult male, was shot, and brought me this morning. This species is not often observed in our neighbourhood, except when the weather is severe, which has not been at all the case this present season. Authors generally describe the smew as feeding on small fish, crustacea, and aquatic insects. The stomach of the present individual, however, was entirely filled with vegetable remains, consisting apparently of some species of *fucus* or *ulva*, mixed with gravels.—The intestines were very long, and the liver remarkably large. The lower part of the œsophagus was greatly dilated, and filled with a fluid substance.

DABCHICK.†

Feb. 22*nd*, 1827.—A DABCHICK was shot to-day. Its stomach contained *dyticidæ* and the remains of other aquatic insects. There was an enlargement of the œsophagus, immediately before the cardiac opening, almost equalling in size half the stomach itself.

RAZOR-BILL.‡

A RAZOR-BILL AUK was picked up alive at Wendy, near Wimpole, in this county, in October 1835,

* *Mergus albellus*, Linn.   † *Podiceps minor*, Lath.
‡ *Alca torda*, Linn.

and is now in the Museum of the Cambridge Philosophical Society.   Thus it appears that birds, ordinarily of the most decided oceanic habits, are occasionally met with inland.   The occurrence of this
species so far from the coast is the more remarkable, from its belonging to a genus in which the
wings are short, and ill-adapted for any extensive
flight.*

GANNET, OR SOLAND GOOSE.†

In this species, we have another instance of an
oceanic bird being sometimes met with very far from
its usual haunts, less extraordinary perhaps than that
above mentioned, from the great powers of flight
which the gannet possesses, and which would enable
it to traverse vast tracts of sea or land in a very short
time.   Two of these birds, both adults, were killed
in Cambridgeshire in the autumn of 1824; one near
Fulbourn on the 11th of October, the other in the
fens between Ely and Southery, about a week afterwards.   Montagu and other authors observe that, in
the autumn, the gannets leave their breeding stations
on the northern coasts of the kingdom, journeying
southward, and that they may be occasionally seen
throughout the winter in every part of the British
Channel, but generally keep far out at sea.   The above
is the only instance I ever heard or met with of their

---

* White mentions an analogous instance of the little auk being
found alive near Alresford, in Hampshire.   *Nat. Hist. of Selborne*, Lett. XXXIX. to Pennant.

† *Sula bassana*, Selb.

occurring inland, and must have been occasioned by
some very peculiar accident. The weather had been
rather stormy and unsettled for a week or two previous, attended by a good deal of wind from S. and
S. E.

BLACK TERN.*

THESE birds occasionally frequent our fens during
the summer months, but not in such abundance as
formerly. Immense flocks, however, appeared in
Bottisham and Swaffham Fens in the summer of
1824, which was a very wet season.—Many of the
specimens which came then under my observation
differed considerably from each other in their
plumage, particularly with respect to the colours of
the head and throat. These parts, which in the
winter are much varied with pure white, generally
become in the breeding-season wholly black, or at
least of a dark ash-colour like the rest of the body :
but in some of these individuals no such alteration
had taken place ; the forehead, space between the
bill and the eyes, throat and fore-part of the neck,
being as white as at other times of the year. Yet
this was on the 8th of July, when the season of
incubation was going on; as was proved by our finding a nest the same day containing two eggs, which
were in a forward state for hatching. This nest was
placed on the ground, and about the size of a saucer,
quite flat, and composed of roots and dry grass,

* *Sterna nigra*, Linn.

K

trodden down so as to be quite firm and compact.
The eggs were barely an inch and a half in length,
of an olive-green colour, thickly spotted and blotched
with deep brown, especially towards the larger end.

CODDY-MODDY GULL.

A SPECIES of gull not unfrequently visits our
meadows and newly ploughed lands the latter part
of the autumn and during winter, provincially called
coddy-moddy.   It appears, from the specimens I
have seen, to be the common gull of English
authors* in its second year's plumage : in its adult
state, I have never known this gull to occur in this
neighbourhood.   Sometimes it is observed as early
as the middle of October.   One, shot on the 22nd of
that month, in 1825, had the head, neck, rump, and
all the under parts white, spotted with light brown;
the back bluish ash-colour; quills dusky, neither
tipped nor anywise spotted with white; wing-coverts
pale brown; tail white, with one broad bar of black
near the extremity; bill of a livid dirty-white at
the base, black at the tip; feet livid, or very pale
flesh-colour.

This bird has retained the name of *coddy-moddy*
in these parts ever since the days of Willughby and
Ray, who notice it in their respective works under
that title as its Cambridgeshire appellation.

* *Larus canus*, Linn.

Dr. Thackeray, Provost of King's College, Cambridge, sends me in a letter the following interesting particulars relating to an individual of the common gull, which he has had for many years alive in his garden, and which is remarkable for having brought up a young duck during the summer of 1844. With this bird the gull seems to have contracted a close intimacy:

" MY DEAR SIR,

" You are aware that the gull in my garden has for several years laid eggs. She had often shewn a disposition to sit, to which I paid little attention, knowing that her eggs must necessarily be unproductive. Last year, at the suggestion of Mr. Yarrell, I placed three duck's eggs in her nest. As it was found that she could not conveniently cover these, one was removed; and she sat constantly on the remaining two till the bill of a duckling appeared from one of them. This she killed in her awkward attempts to extract it, from which I infer that she probably had never bred in a wild state. The next day a young duck was produced, which soon found its way to a fountain distant about twenty yards from the nest. Before she sat, she had been in the habit of roosting on a stone step: on the second or third night after the young duck was hatched, she contrived, by some means or other, to place the duckling by her side on this step. I conceive that this was effected with difficulty; for on the night above referred to, I heard her call, which is different from her common note, after ten o'clock, and saw

K 2

the duck by her side at five o'clock on the following morning. The birds are inseparable, and agree very well; except when the duck, a very powerful bird, is disposed to take a portion of the gull's food. We endeavour to prevent quarrels by feeding the duck with barley, which the gull will not touch; but as he will eat almost any kind of food, and eats much faster than the gull, he is occasionally made to feel the sharpness of her beak. They are generally within a few yards of each other, whether walking or resting. At the latter period, during sunshine, the gull always sits in the sun, and the duck in the shade, even when the ground is covered with snow. There is one remarkable thing with regard to the gull, that whenever she has been handled, either that her wing may be cut, or on any other occasion, she, on being set free, immediately goes to the water and washes herself most carefully, as if she felt contaminated by the touch of any human being.

" I remain, my dear Sir,

" Yours truly,

" GEORGE THACKERAY."

" King's Lodge,
March 22nd, 1845."

OBSERVATIONS ON REPTILES.

COMMON LIZARD.[*]

I HAVE not unfrequently found specimens of the common lizard in a languid quiescent state beneath the bark of felled timber, even during the summer months. Upon a close examination, these have always proved to be individuals which had had their tail fractured, and which appeared to have retired to such places of concealment until the reproduction of it was completed. It is well known, that the tail of this animal is extremely brittle, and that a very slight blow or pressure is sufficient to cause it to separate immediately from the body: no blood issues from the wound, but the severed part continues to move backwards and forwards, and to shew signs of life for a considerable time afterwards. The tail is easily reproduced; and in different individuals which have sustained such an accident, the new one may be observed sprouting, of various lengths, dependent upon the length of time that has elapsed since the fracture. But, though perfect in other respects,

[*] *Zootoca vivipara,* Jacq., Bell.

the new tail never acquires fresh vertebræ ;* and if carefully opened with fine scissars throughout its length, a stoutish nervous chord will be found occupying their place, reaching to the extreme tip, the vertebræ stopping where the original fracture took place.—So common is it for this animal to have its tail injured, that sometimes several may be opened, one after another, apparently having the tail quite perfect, but in which the vertebræ will be found stopping short of the tip by a longer or shorter interval, indicating the extreme portion of it to be of after-growth, in consequence of such accident, as is above alluded to.

———

The number of living young sometimes produced by this reptile is very considerable. I once opened a gravid female, which I found in the fens near here, from which I extracted no less than *ten* young ones, fully formed, and apparently ready for exclusion.† Each was closely coiled up in its own ovum, the coats of which were very thin and membranous; nevertheless, when stretched out, these little ones already measured an inch and an half, though the entire length of the mother scarcely exceeded five inches. This must have been a heavy burden for the poor parent to carry about, the whole cavity of

* This is not offered as a new fact, though perhaps not generally known. I believe Dugés was the first who made the observation.

† Mr. Bell, in his *British Reptiles*, says—"the usual number is from three to six."

whose abdomen seemed nearly occupied by her infant brood.

These reptiles feed on various kinds of insects. The stomach of one that I opened contained wood-lice: another was stuffed entirely with spiders.

COMMON SNAKE.*

SNAKES abound in our fens, where they sometimes attain a large size, occasionally measuring more than four feet in length. When surprised in such situa-tions, they generally betake themselves to the water, in which element they not only swim freely, but have the power of remaining at the bottom for a considerable time without inconvenience. I have sometimes watched one, thus secreting itself from observation amongst the weeds, until my patience was exhausted, and I was forced to leave it, without waiting for its emersion.

The snake is generally first seen abroad about the beginning of April; and on the 22nd of that month, I have found the sexes in copulation: during this act, they are extended side by side in a straight line.

COMMON FROG.†

THE frog, with us, spawns about the middle of March, and the young tadpoles are hatched a month or five weeks afterwards, according to the warmth of

---

* *Natrix torquata*, Flem.   † *Rana temporaria*, Linn.

the season. By the 18th of June, I have observed these to be nearly full-sized, and beginning to acquire forefeet: and towards the end of that month, or the beginning of the next (varying in different years), the young frogs may be seen in great numbers, forsaking the water in which they were bred, and coming on land.

COMMON TOAD.[*]

FROM many years' observations, I find that the toad is invariably a few days later in spawning than the frog. In some seasons, this difference has amounted to more than a fortnight.

Where frogs and toads, at least those in this neighbourhood, pass the winter, I have never satisfactorily ascertained. Much mystery seems to me still to hang over this part of their history, notwithstanding the observations of authors.[†] No sooner is the

---

[*] *Bufo vulgaris,* Flem.

[†] According to Mr. Bell, frogs lie torpid, during winter, " in the mud at the bottom of the water," where they "congregate in multitudes, embracing each other so closely as to appear almost as one continuous mass."—(*Brit. Rept.* p. 89.) The toad, according to the same naturalist, " chooses for its retreat some retired and sheltered hole, a hollow tree, or a space amongst large stones, or some such place, and there remains until the return of spring calls it again into a state of life and activity."—(*Id.* p. 107.) No doubt these statements are true in respect of some particular cases that have been noticed, but, I suspect, are not applicable to all.

Another naturalist has recorded in *The Zoologist* (vol. i. p. 321) a very curious instance, in which he observed a numerous

severity of the winter fairly broken (it will be seen afterwards I speak in reference to former years) than they appear in countless numbers at the bottoms of all our ditches, ponds, and other stagnant waters, where previously there was not one to be seen. This congregating of individuals, which, as is well known, is for the purpose of breeding, may be observed from the middle or end of February (according to the weather), on to April or May. There is a large piece of water at Bottisham Hall which formerly always abounded with toads at this season. Yet though I have often narrowly watched the spot for some days previous to their appearance in the water, I could never detect them in their passage towards it; or, in the idea of their passing the whole winter there, could I observe them going to it at any period of the autumn. What is also noticeable, (for whether they pass the winter in the water or not, they do not pass the entire summer there,) I never could observe any of the old toads coming ashore when

army of some species of toad (thought by him to be distinct and not yet described) in the month of March, and again in autumn, migrating to and fro, from their hibernal quarters, to their breeding localities. This movement, which was over a distance of two miles, was noticed annually for several years in succession; and their winter retreat was discovered to be an old sand-pit, where he found them in the act of burying themselves. The reader is referred to the above journal for a detail of other circumstances connected with this discovery, which is of extreme interest in the history of this reptile. Nevertheless such a migration as this can hardly take place in respect of all toads in other localities, or surely it would have fallen under the notice of other observers before now.

the breeding season was over; their disappearance seeming as mysterious as their appearance in the first instance. Why should the coming of the young broods on land, after quitting the tadpole state, be so generally obvious, and not that of their parents at the expiration of the breeding season? It is probably not so much due to the superior numbers of the former, as to the movement being on their part a more simultaneous one; and the old toads, at whatever season they take to, or quit, the water, must make the passage, I conceive, at different times, or only a few together, and not in large parties, thus to escape observation.

What has also always very much struck me is— the great seeming disproportion between the numbers of toads we observe in stagnant waters during the spring, and the scattered few that are to be found on land at other periods of the year. Here and there one is turned up, or is seen slowly making its way across our garden paths on a summer's evening; but we hardly find them in such plenty as would lead us to suspect the existence of so many in the immediate neighbourhood, as are required to stock our ponds at the above season in the way alluded to. We may infer from this how much there is of life and enjoyment going on about us that we know nothing of: how, among the lower animals, species may abound in certain localities to a degree that the naturalist himself is hardly aware of, from not being sufficiently acquainted with their exact haunts.*

---

* Instances of this kind, among insects, will be familiar to every entomologist; and nothing more strikingly illustrates the

It sometimes happens that toads find their way into situations, from which it is impossible for them to make their exit in order to take to the water during the breeding-season. Such is the case with a small colony I am acquainted with, consisting perhaps of a dozen or twenty individuals, which have for many years had their residence in a damp cellar, to which are underground windows, with a small recess in front of each, the light and air being admitted from above. These underground recesses are covered at top with iron bars, through which toads might easily fall, though they could not get out again by reason of the steepness of the sides. Those I have just alluded to as incarcerated in this manner begin to recover from their winter's sleep at the usual period, and, during the spring, shew great activity, keeping up a perpetual croaking, and sticking to one another's backs, just as they may be seen in the water at this season. The sexual excitement however seems to stop here; and, as far as I have observed, the female never deposits any spawn, which of course would come to nothing without the proper element necessary for its maturation.

I have stated above the circumstance of a large piece of water at Bottisham Hall, which always used

above remark, than the fact of the immense numbers of coleoptera, some usually reputed rare, which have occasionally been taken during floods, collected upon the drifting herbage, and carried by the waters to one spot. See notices of this kind in the *Entomological Transactions,* vol. i. (1812) p. 315; and more recently in the *Zoologist,* vol. i. pp. 116 and 177. There are many other similar ones on record.

formerly to abound with toads in the spring, multitudes being to be seen either crawling at the bottom, or swimming at the surface. It is a curious incident worth recording, that for the last three years not one has been observed at that season in that locality, or in any of the ponds and ditches in the adjoining neighbourhood. At least I have not observed any myself, though constantly on the watch for them; and, on mentioning the circumstance to another, he told me he had noticed the same thing himself. A damp plantation of some extent abuts upon the water, and no clearance or other alteration of the adjacent soil has taken place to explain this phenomenon. *Query*, whether the toads have deserted their old quarters from any cause, or a mortality has taken place among them, so that the few which have escaped its influence are not observed?

NATTER-JACK. *

THIS species of reptile is found in some plenty on Gamlingay Heath, in this county, where it was first noticed by Professor Henslow and myself in August, 1824. Several, which I brought away with me on that occasion, I succeeded in keeping alive during a period of nearly two months. For the first fortnight after their confinement, these animals refused every kind of food that was offered to them;

---

* *Bufo calamita*, Laur.—The account here given of this species is extracted from a paper read by myself to the Cambridge Philosophical Society in 1830, and printed in the third volume of the Transactions of that body.

nor could I perceive that they ate any thing which
happened to fall in their way, though they retained
both their plumpness and activity. At the end of
that period, however, they became more reconciled
to their situation, and readily devoured flies and
other insects that were placed before them, although
it was absolutely necessary that these should be
given them alive : indeed, in no instance could they
be induced to touch their prey, till it began to move,
and to shew signs of preparing to escape. Their
manner of seizing their food was very curious. As
soon as an insect was thrown down into the cage in
which they were kept, the first individual that saw
it immediately pricked up his head, turned quickly
round, and ran towards it till it got within a certain
distance, when it would again stop, crouch down
upon its belly with its hind-legs stretched out, and
gaze at it with all the silent eagerness of a staunch
pointer. In this position it would always remain
till its prey began to move; when, just as the victim
was about to make its escape, it would suddenly
dart out its tongue, and lick it up with a rapidity
too quick for the eye to follow. Sometimes, however, especially if the insect were nimble, it would
follow it about the cage for a considerable time
before it would attempt to secure it, stopping every
now and then to gaze at it, apparently with much
delight, for many seconds together. Nor, in its
endeavour to seize its food, was it always able to
measure its distance with correctness ; often falling
short of its aim, and making two or three fruitless
attempts, before it was finally secured. When, how-

ever, this was once accomplished, the booty was swallowed instantly, excepting when above a certain size; in which case the natter-jack would occasionally remain for ten minutes afterwards with one-half of the insect in its throat, and the other hanging out of its mouth.

The food which these reptiles seemed to relish most, consisted of the smaller species of *diptera* and *hymenoptera*, though they would occasionally take wood-lice and even centipedes. They also ate large quantities of a small red maggot which often abounds in decayed boleti, and any of the lesser *coleoptera* which might happen to stray into their cage. One of them, in a single instance, attacked an ant; but the morsel did not appear to be much relished, for it was no sooner conveyed to its mouth, than rejected again with great haste and trepidation, probably in consequence of the strong acid secreted by these insects. They did not, however, appear to suffer from the stings of the smaller bees and ichneumons, which were repeatedly swallowed with impunity.

The natter-jack is a much more lively animal than the common toad; and when in search of food, or following its prey, shews great alertness. When full fed, or from other causes inactive, the above individuals would conceal themselves in a sod of turf, which was always kept in their cage. They also occasionally delighted much in a pan of water, in which they would float motionless for half an hour together, having all their legs stretched out, and no part of their body except their head above the surface. But the great distinguishing habit of this species is its mode of progression. Unlike the frog, which advances by regular leaps; and the toad, whose pace is seldom exerted beyond that of a slow crawl, —the natter-jack has a kind of shuffling run, which is seen to most advantage when it is following its prey, and by which means it is enabled, when in full health and activity, to get over its ground with considerable quickness.*

* A yet more interesting addition than the natter-jack to the reptiles of Cambridgeshire has been made lately in the instance of the edible frog (*Rana esculenta*, Linn.), which has been met with in Foulmire Fen, in some abundance. See *The Zoologist*, pp. 393, 467, 677, and 727, for different notices relating to this new and important discovery.

OBSERVATIONS ON FISHES.

RUFFE.[*]

Two specimens of a singular variety, or rather monstrosity, of this fish are preserved in the Museum of the Cambridge Philosophical Society. In these individuals, the head is, as it were, truncated in front of the eyes; the profile being exactly vertical, and presenting a remarkable bull-headed appearance. The eyes are very large and prominent, and project more in front than behind, which causes them to have a somewhat backward direction. The mouth is small and contracted, with the lower jaw longest. In all other respects they resemble the common kind.—Some of these peculiarities are probably due to a malformation of the intermaxillary bones, such as may occasionally be met with in other species.[†]

The above specimens were taken in the Cam, in which river this species of fish is far from uncommon. With us it is generally called a *pope*.

[*] *Acerina vulgaris*, Cuv. et Val.

[†] See a vignette in *Yarrell's British Fishes*, vol. i. p. 110, representing the head of a sea-bream, in which the intermaxillary bones are entirely wanting.

GREAT WEEVER.*

I AM not aware that this fish, which is met with occasionally on several parts of the British coast, has ever been recorded as entering rivers. Yet a few years back a full-sized specimen, now preserved in the Museum of the Cambridge Philosophical Society, was taken in the Ouse a little below Little-port, at a distance from the sea of at least twenty miles. The Ouse is not a tidal river; the sea-water in general being kept out by the doors of Denver sluice near Downham, and compelled to ascend the Hundred-foot river instead. This renders the circumstance more unlooked-for. It appears, however, that on this occasion the doors of the sluice by some accident were prevented from closing so quickly as usual, in consequence of which a considerable quantity of sea-water was admitted into the river, and with it such fish as happened to be present in the water just at the time. Still the great weever, in this instance, must have freely ascended from the sea as far as the sluice; and its entering a tidal river at all, is, as far as I know, a new circumstance in its history.

RIVER BULLHEAD, OR MILLER'S THUMB.†

THE miller's thumb, when full-grown, is a voracious fish, and not always content with " the larvæ

---

* *Trachinus draco,* Linn.   † *Cottus gobio,* Linn.

of water-insects, ova," or even " fry," said to be its usual food.   In the Museum of the Cambridge Philosophical Society is a large specimen, measuring four inches and a half in length, which was found at Chesterton sluice, lying dead on the surface of the water, having been choked in the attempt to swallow one of its own species, and more than half its own size.   Its prey was still sticking out of its mouth when discovered.   Both individuals are preserved in spirit exactly as they were found.

TEN-SPINED STICKLEBACK.*

THIS is one of the smallest of our British fishes, and rarely exceeds the length of two inches, being usually, indeed, much less.   When found, however, in stagnant water, it sometimes grows considerably larger.   In the Museum of the Cambridge Philosophical Society are some very large specimens from some pits near Madingley in this county,—the largest measuring full two inches and three-quarters in length.

GOLD-FISH.†

IT is well known that gold-fish, when kept in a state of domestication and confinement, are subject to great variation of form and character.   Occasionally the dorsal fin is extremely small, or entirely wanting; in some, the caudal acquires a preter-

---

* *Gasterosteus pungitius*, Linn.   † *Cyprinus auratus*, Linn.

natural size, or is divided into several lobes; in
others the eyes protrude, and are very much swollen.
In the Cambridge Museum, however, is a specimen
so extremely monstrous, and whimsical in shape and
general appearance, as to deserve a more detailed
notice. This individual came direct from China,
and formed part of a collection made in that country
by the late Rev. George Vachell, Chaplain to the
British Factory at Canton, and presented by him to
the Cambridge Philosophical Society.

Its chief peculiarity is its form, which almost
approaches to globular, at least below, like that of
the genus *Diodon*. This arises from the extreme
rotundity and protuberance of the abdomen; the
back being at the same time slightly arched, more
especially behind, and the fleshy part of the tail as
if entirely cut away; the caudal fin being set on a
little behind the dorsal, and immediately above the
anal. This gives the posterior part of the body an
unusually blunt and truncated appearance. The
caudal itself is very large, double, but the two portions united at top, and folding double, in a vertical
direction, like the tail-feathers of the common fowl.
The dorsal is single, and has nothing very remarkable
in it. The anal is double like the caudal, but much
smaller than that fin, and so directly beneath it as
almost to be concealed within its folds: also, from
the extreme convexity of the abdomen, the line of
its base, where the rays unite with the body, is
actually vertical. The pectorals and ventrals are
much as in other examples, though rather longer,
and more pointed, than usual. There is nothing

else very peculiar in the character of this fish, excepting the eyes, which are enormously large and
protuberant. The length of the head and body,
measured from the end of the snout to the setting
on of the caudal, is three inches and a half. The
greatest depth is two inches and a half; being five-
sevenths of the length instead of about one-third, as
in a fish of this species of ordinary proportions.

POMERANIAN BREAM.*

THE occasional occurrence of single individuals
of species in situations where they are not known
generally to inhabit, especially species that are rare,
and hardly considered as abundant anywhere in the
country, is sometimes a very puzzling circumstance.
Birds, which have the power of transporting themselves by wing to great distances, may accidentally
wander to many localities in which they would be
accounted strangers. Quadrupeds, and a few other
land animals, though more limited in the possible
extent of their range, may still, from some casualty
or other, be carried far beyond their usual beat : the
smaller ones, like insects, may escape observation for
some time where they are really plentiful. But the
difficulty is much increased in the case of the larger
aquatic animals residing in waters which have no
communication with the sea, or rivers. And such is
the case in the instance to which these remarks allude. In June, 1839, a single specimen of the Pome-

---

* *Abramis buggenhagii*, Yarr. *Supplement to Brit. Fishes*, p. 39.

ranian bream, a species of fish which had only been discovered, for the first time, in this country, three years previous, and then in but one locality in England, was taken in the water in the park at Bottisham Hall. No other individual had occurred before, nor has it been met with since, though often sought for. Mr. Yarrell first recorded this species as British, having received a specimen from Dagenham Breach in Essex, which water communicates with the Thames, and from thence probably the species was introduced into it.* In the instance above mentioned, the water is supplied from a running stream, which rises at a spring issuing out from underneath the chalk, about a mile up. No fish can possibly find their way of their own accord into this piece of water, except such as come with the stream that supplies it. And I can only attribute the occurrence of this bream in it to the circumstance of its having been brought from the river, when very small, along with other small fish of the common kinds, which have occasionally in times back been supplied to the water in question to serve as food for the pike which are preserved therein. Yet it is observable that I never knew of this species of bream being taken in the river, though repeatedly fished; and I have often been present myself in former years, when it has been dragged by nets, with the view of ascertaining what species it produced.

The above specimen of the Pomeranian bream is

* Mr. Yarrell informs me that he has since received a specimen of the Pomeranian bream from Wolverhampton in Staffordshire.

now in the Museum of the Cambridge Philosophical
Society. It measures 12 inches in length, with a
depth of 3 inches, 6½ lines; the greatest thickness
being 1 inch, 5 lines. It is distinguished from both
the two more common kinds of bream,* by its greater
thickness in proportion to its other measurements;
and its much shorter anal fin, with fewer rays: and
from the yellow or carp bream by its much larger
scales, giving a smaller number both in the length
and depth. In this individual the number of scales
in the lateral line is about 50, and in the depth 16;
of these last 10½ are above the lateral line, and 5½ below it. The fin-ray formula is as follows: D. 11, the
last double; A. 17, the last double; C. 19; P. 17;
V. 9. The colours of this fish, when fresh out of
the water, were as follows: upper part of the head
and back dark olivaceous brown, becoming lighter on
the sides, and passing into silvery on the belly; a
faint golden hue, however, everywhere pervading:
dorsal, anal, and caudal fins dusky, tinged with pale
red; pectorals and ventrals the same, but paler, or
with the reddish tint rather prevailing over the
dusky: cheeks and gill-covers silvery, inclining to
golden yellow: irides silvery.

RED-EYE OR SHALLOW.†

March 12, 1831.—THE *shallow*, as it is called
by the fishermen of our river, though more com-

* The yellow, or carp bream (*Abramis brama*, Cuv.), and the
white bream, sometimes called *bream-flat* by the fishermen about
Ely, (*A. blicca*, Cuv.)   † *Leuciscus erythrophthalmus*, Cuv.

monly known by the name of *red-eye*, is peculiarly
abundant in Reche Lode, a navigable cut from the
Cam, near here.   Females brought from that locality to-day were found, when opened, to be full of
roe, though far from ripe : the exact spawning time
of this species I have not been able to ascertain ; I
believe it, however, to occur about the third or fourth
week in April.

This fish is remarkably distinguished from its congeners by the bright vermilion colour of the ventral,
anal, and caudal fins ; and by a peculiar faint golden
or brassy lustre, somewhat resembling the tint of a
bad shilling, pervading the whole body.

GROUNDLING.*

THE groundling is a small fish, apparently very
little known in other parts of England,† but far
from uncommon in our river, and in ponds and
ditches supplied from it.   I have frequently taken
it in the Cam and Reche Lode, as well as in the
Ouse below Ely, and in fish-ponds adjacent to the
river at this last place.   From its keeping, however,
very much in the mud, it is not easily obtained,
except by a net that scrapes close to the bottom.
Its partiality, indeed, for mud, and habit of residing
occasionally even in the thick sediment of stagnant
waters, would seem to distinguish it from its more

---

* *Cobitis tænia*, Linn.

† Mr. Yarrell, in his *British Fishes*, mentions no other localities
for this species except those recorded by Berkenhout and Turton,
and the one I furnished him with.

common ally, the loach, which, I believe, is more attached to clear running streams with a gravelly bottom. I have kept groundlings alive in a large jar of water for several days, but they remained wholly at the bottom, in the mud I had introduced into the jar, and seldom or never quitted it, except under the influence of alarm, when they would suddenly dart about for a few moments, and then return to their former place of shelter. The small forked spine immediately beneath the eye in this species, easily felt by the finger when passed from below in an upward direction, if not readily seen, at once characterizes it; the body is also much more compressed, and the barbules shorter, than in the loach, the only other species with which it can be confounded.*

* There appear to be two well-marked varieties of the groundling, (if they be not distinct species,) which have hitherto been undistinguished by ichthyologists. The difference between them is as follows:

(1.) One, which I would denominate the *large-headed groundling*, is thicker both in the head and body (the length of the two specimens compared being the same) than the other: the thickness of the head in the region of the gills is equal to that of the body, and both are about two-thirds of the depth: the head is also larger; with the profile very convex in front of the eyes, whence it falls vertically to the lips; the snout, consequently, is obtuse, with the mouth at bottom, and the lower surface of the head in nearly the same horizontal line with that of the abdomen: the barbules are rather longer than in the other kind; the suborbital spine, on the other hand, less developed: the eyes are high in the cheeks, the space between narrow, but not elevated into a ridge.

(2.) The second, which, for distinction's sake, I call the *small-headed groundling*, is more compressed than the last, especially about the head, which is narrower than the body, and which

PIKE.*

SOME years ago a pike was observed lying dead on the water in Bottisham Park, having been choked in its attempt to swallow one of its own kind nearly as big as itself, which was sticking out of its mouth. A similar instance of voracity in the miller's thumb has been already mentioned in a former part of this work.†

Mr. Selby also tells me, in proof of the voracity of this species of fish when urged by hunger, that he has at various times hooked pike, when fishing with minnows as bait, and had his tackle, and a considerable part of the line carried off, yet has taken the

appears, in the region of the gills, as if pinched in by the fingers: the greatest thickness of the body is scarcely more than half the depth: the head is smaller, and of a different form, somewhat approaching to triangular, when viewed laterally: the snout, in advance of the eyes, is not so obtuse; the profile less curved, and never becoming vertical; the mouth is not so low down, and the lower part of the head ascends obliquely to meet it: the barbules are a trifle shorter; the suborbital spine larger and more conspicuous: the eyes are very high, the intervening space contracted into a narrow elevated ridge.

The fins, and fin-ray formula, as well as the colours, are the same in both.

The habits also of these two kinds of groundling are, as far as I have observed, the same. The largest specimens I have obtained were the large-headed variety, and measured three inches four lines in length: these were taken in fish-ponds at Ely.

* *Esox lucius*, Linn. † See p. 210.

same fish again in a *quarter of an hour,* and thus recovered his tackle.

———

Pike, even of a fair size, often find their way into the turf-pits in our fens, when the fens are partially flooded. When the waters subside, they remain imprisoned in the pits, and in such situations are occasionally taken by the fen-men, employed in digging turf, in a peculiar manner. They stir up the mud at the bottom of the pits with their tools, till the water, of which there is often not much, becomes so thick that the fish is almost suffocated; it then floats at the top, and is easily secured.

SMELT.*

April 10, 1824.—A QUANTITY of smelts were sent us to-day from the Hundred-foot River, where they are annually taken about this period of the spring. At Mepal, and other places higher up, large numbers are caught for the table, from whence they are distributed about the county. It is stated by the persons who take them in that neighbourhood, that these fish are then going up the river.† Those we received to-day were finer and larger than any I

* *Osmerus eperlanus,* Flem.

† Smelts are generally supposed to ascend rivers from the sea in August, or during some part of the autumn. On mentioning the above circumstance to Mr. Yarrell, he said it was a new fact to him that they should be only going up the river in March; but

remember to have seen before : they were exceedingly plump, and full of spawn, which had not yet been shed.

These fish, when fresh, have a peculiar perfume-like odour (whence their English name of *smelt*), which affects the hand for a considerable time after handling them, and which is not to be got rid of without repeated washings.

March 21, 1826. — Smelts received to-day, as in former years, about this time, from the Hundred-foot River, but earlier than usual.* I am told, however, that they are sometimes taken there as early as the 10th of this month. The people continue fishing for them till about the 15th of April. Many of those sent us on this occasion had their gills much infested by small worms. These worms were of a whitish colour, about three inches in length, somewhat filiform in the middle, and about half the thickness of a small crow-quill, slightly tapering towards each extremity. The condition of the fish did not appear to be impaired by the presence of these parasites.†

he thought that distance from the estuary, in this instance, might account perhaps for their late appearance.

* Persons resident near rivers which are annually visited by these fish, or any other species of the *Salmonidæ*, would do well to notice the exact periods of their ascent and descent, from and to the sea, in connection with the character of the season. Both, no doubt, vary much in different years, according as the weather may be more or less seasonable. This is a point which will be further adverted to in a subsequent part of this work.

† Dr. Bellingham, well known for his researches on the *entozoa*, and to whom I lately sent specimens of this parasite, was kind

EEL-POUT OR BURBOT.*

EEL-POUTS, or burbots, as they are more generally called in other parts of England, are common in the Cam, and the navigable cuts communicating with that river. In Reche Lode they are frequently taken, where they sometimes attain a considerable size. One brought to me from thence in May, 1829, weighed three pounds four ounces and a half: its length was two feet all but half an inch; and its girth exactly half its length, or nearly a foot. This individual was looked upon by the fishermen as rather larger than those which usually occur.

The eel-pout is a fish hardly known in many rivers; and this circumstance, connected with its appearance and shape, which are a little forbidding to those who see it for the first time, occasions with some a prejudice against eating it. It is, however, a very delicate fish at table; the flesh firm and white, and of agreeable flavour, most resembling that of the common eel.

A singularly coloured variety of the eel-pout was

enough to inform me that it was an *Ascaris,* and probably the *A. obtusocaudata,* which inhabits the stomach of the genus *Salmo* ;† but that, without inspecting male and female specimens, he was not able to identify the species positively. He observed further, that the species of *Ascaris* which reside in the stomachs of fish, not unfrequently make their way thence to the gills, as in the instance above spoken of.

* *Lota vulgaris,* Jen.

---

† *Ann. and Mag. of Nat. Hist.* vol. xiii. p. 172.

taken in the river Cam, at Clay Hithe, in March
of the last year (1845), and is now preserved in the
Museum of the Cambridge Philosophical Society.
The whole fish, when first taken out of the water,
was of a rich golden-yellow, with the exception of a
few black spots on the fins. One of these spots was
in front of the first dorsal; two were on the second
dorsal, one a little beyond the middle, the other at
the extremity of that fin; besides which there were
two or three dark stains on the caudal. The length
of this fish rather exceeded eight inches.

EEL.

MANY years back, two eels of an enormous size
were taken in a drain near Wisbeach, weighing together not less than fifty pounds; the larger one twenty-
eight, the smaller twenty-two pounds. The length
of each was upwards of six feet, and the girth equalling that of a man's leg. These eels were of the
sharp-nosed kind,* and probably the largest of which
any record exists. Their stuffed skins were long
exhibited at a fishmonger's shop in Cambridge, where
I took Mr. Yarrell to see them in the summer of
1830, who assisted me in the determination of the
species. Since then the shop has been removed, and
I am ignorant what has become of them, or whether
even they are still in existence. It is worthy of note
that these eels were taken on the occasion of clean-
ing the drain out, and that no other fish of any kind,

* *Anguilla acutirostris*, Yarr.

according to the report I received, were found with them.

Feb. 10, 1844.—A large eel was found to-day in the stew-hole at Bottisham Hall, deeply imbedded in the mud. The weather this month has been very severe.

The occurrence of eels during the winter is not very unfrequent. I remember, after the breaking up of the frost in the hard winters of 1814-15 and 1829-30, the surface of the piece of water in Bottisham Park was covered for a time with dead eels of various sizes, which had been killed by the severity of the season.*   This circumstance proves what in-

---

* This fact was mentioned by me to Mr. Yarrell, who has recorded it in his " Observations on Eels," published in *Jesse's Gleanings in Natural History*, 2nd series, p. 71.

One object of Mr. Yarrell in the "Observations" just alluded to, was to prove that eels are oviparous, like other fishes, and not viviparous, as has been frequently supposed. I have no new evidence myself to adduce in illustration of this part of their history, and indeed, I may say, I am almost thoroughly satisfied with the correctness of Mr. Yarrell's views. Yet it may be well to take this opportunity of calling the attention of naturalists to a communication made to the Royal Academy of Sciences at Paris, in Feb. 1839, on the subject of the generation of eels, and the author of which seems to think that he has established the fact of their being viviparous after all. One of his observations in proof of this statement rests on the authority of a countryman who is said to have fished up a large eel on the 20th of March, which he took into his house, and put into a large hollow dish, covering it with another, as he was obliged to return to his work in the fields immediately. What was his astonishment, on coming in again in the evening, and raising the upper dish to get at his eel, to find it surrounded with, it may have been, two hundred young ones,

deed is now, I believe, generally allowed; that though, for the most part, these fish may migrate to the sea before winter commences, many do not;

from an inch and a half to two inches in length, about the thickness of threads, and nearly white!

M. Joannis, who is the author of the communication above alluded to, says, that, on hearing this circumstance, which appeared to him so interesting, and decisive of the question as to eels being viviparous, he pressed the man with numerous questions, the result of which was as follows:

It appears that, the moment the man first perceived the circumstance, the eel was still in the act of giving birth to its young, for he found one which was only half excluded. A small quantity of glairy matter was at the bottom of the dish, but very little; the young ones which were already born beginning to climb, with a serpentine motion, the sides of the large dish; some were glued to it by the posterior part of their body, raising their heads in a convulsive manner; others were dead; others were in a state of agitation, and especially at the bottom of the dish. Their two eyes were observed very distinctly, resembling two large black points. In general, it was remarked that the little ones which crept along the sides of the dish were much fettered in their movements by a gluey matter, with which their bodies were covered, and which caused them more or less to adhere.

I have given the above just as it is stated in a brief notice of this memoir published in *L'Institut*, vol. vii. (1839), p. 67. I am far from thinking that the evidence is conclusive in respect of the point sought to be established. It is to be regretted that the author did not see the young eels himself (which had been thrown away by the man as good for nothing), and thereby prove that they were not mere intestinal worms, as some of the reputed young eels, found in other cases in the body of the parent fish itself, have been conjectured to be. I insert the statement however, for the interest and information of those in whose way the above Journal may not fall, and in order to excite further inquiry on the subject.

especially such as inhabit ponds, and other pieces of water, from which there is no outlet.

STURGEON.[*]

STURGEONS, I believe, are seldom found but in the larger rivers; and then, according to Mr. Yarrell, "most frequently in the estuaries, or but a short distance up."—Every instance of their occurrence, therefore, under other circumstances is interesting; and I may mention the capture of one in the Ouse, near Ely Bridge, on the 18th of June, 1816. This must be at least thirty-five miles from the sea. The weight of this fish was 112 pounds exactly. Its entire length was six feet seven inches: from the extremity of the snout to the insertion of the pectoral fin measured eighteen inches and a half: from the end of the tail to the posterior part of the dorsal fin, nineteen inches. These measurements were taken by myself from the stuffed skin, which is still preserved at Ely. The weight given above is on the authority of the person into whose hands it came when first taken.

[*] *Acipenser sturio,* Linn.

OBSERVATIONS ON INSECTS.*

MISCELLANEOUS.

THE first insects seen on wing in the spring in any considerable numbers are generally the smaller coleoptera. These belong principally to the families of *Aphodiidæ, Curculionidæ,* and *Staphylinidæ.*— Swarms of minute gnats, indeed, as long since observed by White, may be observed playing about the winter through, if the weather be mild and the air still. There are also several other insects of different orders, which may occasionally be noticed here and there, coming abroad to enjoy a casual fine day. But those mentioned above are the first that occur in any plenty; and their appearance is generally reserved for one of those warm spring days, which sometimes occur prematurely in the month of February, when the temperature suddenly rises several degrees, and the air acquires a mild softness, so agreeable to the feelings after the winter's cold hardly yet over, and perhaps returning the day after

---

* The term is here used in a larger sense than that in which it is generally used by entomologists.

in all its severity. Such a day is generally marked too by the appearance on wing of the hive and humble bee, and a few individuals of the yellow* and tortoise-shell butterflies.† These are, as it were, so many *avant-couriers* to announce the approach of that season, which is no less welcomed by man than by the various tribes of animated beings which it recalls to life and activity.

————

Wet is probably more prejudicial to insects than cold; and for this reason mild winters, which are generally attended with much wet, are, I apprehend, more destructive to them than severe ones. Cold renders them torpid; but when once they are brought into this state, it matters little whether the temperature sink a few degrees lower or not, or how long they remain thus inactive. Whereas constant rains percolate into their most retired hybernacula, and drown them in large numbers, whilst they are yet too inert to make their escape. It would be worth inquiring, though I am not aware of anything on record to shew, whether collectors of insects find their harvest in summer depending at all upon the character of the preceding winter, or at least upon its having been wet or dry.

————

White has noticed a "humming in the air," like that of bees, which might be heard upon the highest

————

* *Gonepteryx rhamni*, Steph.   † *Vanessa urticæ.*

part of his down in hot summer days; and with respect to the cause of which he could not satisfy himself, as not one insect was to be seen. He says, " any person would suppose that a large swarm of bees was in motion, and playing about over his head." *—This noise, in fact, may be heard almost everywhere during the height of summer; at least on bright hot days, when the wind is still. It is certainly deceptive; but I have not the least doubt of its proceeding from no particular swarms at any given spot, but from the accumulated multitudes of insects, large and small, on wing together, filling the air in every direction, and causing it to resound with their numbers. Most of these are probably of the dipterous kind, and are not only too small to be distinctly perceived whilst rapidly in motion, but fly at a considerable height in the atmosphere proportionable to the warmth and fineness of the weather.

The occurrence of this universal humming is not more amusing (as White speaks of it) than instructive. It impresses us with an idea of the fulness of all nature. We are greatly struck with the extent to which life teems in the summer season, when, wherever we bend our steps, we thus carry with us such irrefragable proofs of its existence.— The countless myriads about us, though not visible to the eye, are made obvious to the ear; and if we have any of those feelings within us which best become the true naturalist, we shall be irresistibly led to reflect on the immensity of that Being from whom

* *Naturalists' Calendar*, p. 99.

all life comes, and who bestows a capacity for enjoying his gifts on the lowest and smallest of his creatures, as well as on the highest and most considerable.*

———

I have been a collector of insects, more or less, all my life; and I have been often struck with the occurrence of some species, occasionally rare ones, in odd situations, or under certain particular circumstances, and no other. Thus, a small aquatic beetle,

———

* If this occurrence is so calculated to impress the mind of the natural observer in this country, we may imagine what it must be in tropical regions, where life is so infinitely more abundant. Humboldt has a passage on this subject, which, as bearing closely upon the phenomenon above alluded to, I will here transcribe :

"How vivid" (he says) "is the impression produced by the calm of nature, at noon, in these burning climates ! The beasts of the forests retire to the thickets ; the birds hide themselves beneath the foliage of the trees, or in the crevices of the rocks. Yet, amid this apparent silence, when we lend an attentive ear to the most feeble sounds transmitted by the air, we hear a dull vibration, a continual murmur, a hum of insects, that fill, if we may use the expression, all the lower strata of the air. Nothing is better fitted to make man feel the extent and power of organic life. Myriads of insects creep upon the soil, and flutter round the plants parched by the ardour of the sun. A confused noise issues from every bush, from the decayed trunks of trees, from the clefts of the rock, and from the ground undermined by the lizards, millepedes, and *cæcilias*. These are so many voices proclaiming to us that all nature breathes ; and that, under a thousand different forms, life is diffused throughout the cracked and dusty soil, as well as in the bosom of the waters, and in the air that circulates around us."—*Pers. Narr.* vol. iv. p. 505.

only discovered a few years ago, (*Hydroporus ferrugineus*, Rudd,*) not unfrequently appears in the water-jugs used in the bed-rooms in my house, and which are filled from a pump in the yard. Yet I never took this species in any of the adjoining streams, or observed it elsewhere in Cambridgeshire. The *Broscus cephalotes* of authors,† which is always represented in books as a maritime species, and which is found plentifully on many parts of the coast under marine rejectamenta, I once took under stones on the Devil's Ditch, a perfectly dry spot, and at least forty miles from the sea. The *Cychrus rostratus* ‡ I never took, excepting in a common cock-roach trap, when placed in certain out-buildings, but by that means have captured several specimens. The *Apis conica*, L.§ I have sometimes found drowned in my rain-gauge, but never saw on wing in the garden in my life, or took alive within sixteen miles of this place.—Many other instances might be mentioned analogous to the above; and probably many similar ones have occurred to every collector.

———

Novemb. 1845. — I never remember in any year such a dearth of insects as during the past season. Having been much engaged in other ways, and not

* *Stephens's Man. of Brit. Coleopt.* p. 68.
† *Id.* p. 34.   ‡ *Id.* p. 13.
§ *Kirby, Mon. Ap. Angl.* vol. ii. p. 224.   It is the *Cælioxys conica* of Latreille.

having gone to seek them, I speak not in reference to rare and interesting species, but of the commoner kinds, which usually obtrude themselves upon us, go where we will. Many of these, which in ordinary seasons appear at stated times with the greatest regularity, and of which I am in the habit of recording the first appearance in my journal, have not shewn themselves in a single instance. Some plantations close at hand, and through which I constantly pass, and which usually abound with several common *Lepidoptera*, as well as species of the other orders, have been as it were deserted. Such common insects as the scorpion-fly (*Panorpa communis*), and the two horse-flies, *Hæmatopota pluvialis* and *Chrysops cæcutiens*, have been wholly wanting. Scarcely an *Amara*, a *Harpalus*, a *Pœcilus*, or a *Silpha*, have crossed my path in my walks on any occasion. Not one admiral butterfly has been seen in the garden, nor one of the small coppers (*Lycæna phlæas*) in the meadow adjoining : very few even of the common blue there; or of the meadow brown (*Hipparchia janira*) and the great heath (*H. tithonus*) in the hedges. No *Noctuidæ* have entered the rooms after dusk : no bottles have been required upon the fruit-trees to catch the wasps, of which not a score have been seen the whole summer; and hardly any of the great harry-long-legs (*Tipula oleracea*), generally so plentiful, have appeared in the field before the house. The same scarcity of insects has continued through the autumn. The common dragon-fly (*Libellula vulgata*), which generally swarms in August and September near some water in the immediate

neighbourhood, has not shewn itself once. Not
once has been heard the shrill cry of the *Acrida viri-
dissima,* with which the hedges about here resound
in some seasons; and but very seldom the crinking
of the common grasshopper. It is also remarkable
that the few grasshoppers which were observed in
the meadows never seemed to arrive at their full
development, and to acquire wings; but they remained apparently in the pupa state, till, from the
advancement of the season, they ceased to be observed at all. I may add, lastly, that but few *Diptera*
have appeared on the windows throughout the year.
Even the well-known hum of the common blue-bottle
has been but seldom heard in the apartments; and,
as the year declines, the number of flies now entering, and congregating in the house, is without exception far below the usual number on any previous
occasion. At the same time it may be noted, that
white butterflies were in tolerable plenty during the
summer, and the caterpillars of these insects are
now swarming upon the cabbages.*

* The summer of 1845 was very wet and cold, generally speaking, with a sky almost constantly clouded, the consequence of
which was a great defect of both light and heat. The season will
long be remembered on account of the epidemic which prevailed
so universally, both here and abroad, in the crops of potatoes,
causing rottenness in the roots. This disease proved fatal to
nearly entire crops of these plants in many parts of the country.
It was supposed to have had its origin in the peculiar character of
the weather during July and August.

The greasing of insects in cabinets is a circumstance well known to every collector. I am not aware, however, of its having been observed, that the *males*, at least among the *Lepidoptera*, grease much more frequently than the females. Such at any rate is the case in my own collection, in which, in the instance of the following species more especially, I find the male specimens alone affected: *Smerinthus ocellatus, Sphinx ligustri, Macroglossa stellatarum, Hepialus humuli, Zeuzera æsculi, Cossus ligniperda, Cerura vinula, Saturnia pavonia minor, Lasiocampa quercus, odonestis potatoria, Gastropacha quercifolia, Dasychira fascelina, Leucoma salicis,* and *Spilosoma lubricipeda.*

I think also, I observe, among the Coleoptera, that the timber-eating beetles, such as the *Lucani* and the *Cerambycidæ,* grease more (contrary to what one might have expected) than those which, in the larva state, feed on animal substances. But possibly a more extended inspection of other cabinets might prove this last remark not generally confirmed.

———

It was once remarked to me by an eminent philosopher, who thought nothing in nature beneath his notice, that, having had the curiosity to analyze the excrement of the common silkworm, he was much struck with finding how little change had taken place, in the portions of the leaves which these caterpillars had devoured, by the process of digestion. This induced me to examine the excrement of several species myself. The dung of a full-grown larva of the *Sphinx atropos,* (which I selected on account of its size,) when placed in cold water, almost immediately resolved itself into three smaller masses in no way connected with each other. Each of these masses subsequently fell to pieces, and was found to consist of small fragments of leaves, not larger than would be taken in at a mouthful, and of irregular as well as variable shape, though mostly approaching the form of a spherical triangle: the whole of these fragments, however, in the case of each of the three smaller masses, were enveloped in an extremely delicate membrane, somewhat resembling a cobweb, which was much tangled and folded, and very easily ruptured, letting the small undigested portions of leaf escape. These fragments of leaves were very little altered, and appeared as if they had merely been subjected to compression in order to extract the nutritious particles: under the microscope portions of the ribs were still distinctly visible, as well as the superficial hairs. It would seem, therefore, that there is no solvent power in the stomach or intestines of the caterpillar, over its food, further than as respects the parenchyma of the leaves they devour, which is alone the source of nourishment. Those caterpillars, therefore, which, as most do, devour the whole substance of the leaf, and the larvæ of the subcutaneous *Tineæ,* which never touch the cuticle, are in reality supported in the same way; the only difference being that the former do not separate the digestible from the more indigestible parts in the act of feeding, swallowing the whole indiscriminately, and afterwards voiding that which has no nourishment in it;—whereas the latter make the selection in the first instance, burrowing, as their small size enables them to do, between the upper and lower cuticle, and swallowing the pulp alone.

CARABUS VIOLACEUS. *

SOME years back, in the month of July, observing a remarkably fine specimen of this insect, I made an endeavour to lay hold of it. It escaped between my fingers in the first instance; but shortly afterwards turning upon its back (accidentally as I conceived), and not being apparently able to recover itself readily, I took the opportunity of renewing my attempt to seize it, as it lay struggling. I had hardly secured it, my arm being nearly at full length, and the insect with its anal extremity towards me, before I felt a sensation as if struck across the face with a stinging-nettle, followed by the most intolerably fœtid odour that can be conceived. There is no question but this proceeded from some acrid fluid or vapour ejected by the insect *per anum*, on to my nose and lips. The irritation which ensued was very considerable, and gave rise to a number of little blisters on many parts of my face, which, though washed immediately with water, continued painful for several hours after.

Many of our indigenous *Carabidæ*, when handled, emit from their mouths, by way of defence, a brown fluid, of an acrid caustic nature, highly nauseous to

* *Stephens's Man.* p. 15.

the smell; but I was not before aware that any of them had the means of ejecting this, or any similar liquid, from behind, much less with the force exerted in the above instance. Perhaps this last may be a peculiar substance, of a different quality from the first, secreted in less abundance, and offering a mode of defence only resorted to under particular circumstances, or in the failure of those means of escape which the insect more ordinarily adopts.*

OMASEUS ATERRIMUS.†

In the year 1826, this local insect occurred, in the month of April, in considerable plenty in Bottisham Fen. Specimens continued to be observed till the middle of June. Since then, however, I have but rarely noticed it. All those I found were observed crawling on the bare and wet mud at the edges of the turf-pits. Some were basking in the sun, in a state of quiescence, and were not immediately obvious; but by stamping on the ground with the foot, which gave a tremor to the loose,

* Since writing the above, I find a notice of this habit of the *Carabidæ* by Mr. Holme, in *The Zoologist*, p. 339.

He says,—"I never saw, in any English work on Natural History, any notice of the power possessed by the *Carabi* of ejecting an acrid fluid *à posteriori* with considerable force to a distance of six or eight inches, and generally so well directed as to strike the captor in the eye. This has not escaped the notice of the Continental entomologists, and the incident quoted by Kirby and Spence (vol. ii. p. 244, 5th edit.) is probably referable to it."

† *Steph. Man.* p. 32. This insect is figured in *Curtis's British Entomology*, pl. xv.

boggy soil, they were roused to activity, and the eye caught them directly.

SINONENDRON CYLINDRICUM.*

April 15th, 1826.—To-day we found an old decayed pollard ash completely bored through and through by the *Sinonendron cylindricum*. Immense quantities, both of the perfect insect and their larvæ, were found buried in the wood. It is singular that this species should occur in such profusion in this instance, being one which I have not before observed in this neighbourhood.

COCKCHAFFER.

The larger cockchaffer † is so uncommon in this neighbourhood, that during the last ten years I have hardly noticed as many specimens. I was once inclined to attribute this to the fact of there being an extensive rookery close by, the individuals of which must greatly assist in keeping down these destructive insects, when in the grub or larva state. —It is observable, however, that the smaller cockchaffer,‡ or midsummer dor, as it is more commonly called with us, abounds; and one might have supposed that this species, which the rook is equally fond of, would have been kept down as well.—The

* *Steph. Man.* p. 155.
† *Melolontha vulgaris,* Stephens's Man. p. 168.
‡ *Rhisotrogus solstitialis,* Id. p. 168.

above circumstance, therefore, is due probably to some other cause.

White has noticed the punctuality with which the small midsummer chaffer comes out every year.* He says, it first appears about June 26; which accords exactly with the date of its appearance here. A mild summer's evening, about the end of that month, brings them out by hundreds, and they may then be seen flying about the garden in all directions. Cats are very fond of them, and devour great numbers, which they catch by springing up at them as they wheel over their heads. Very many of these insects, I observe, fall down the chimneys into the rooms below.

It is remarkable, that though this insect abounds to such a degree in my garden every midsummer, yet the pasture in front of my house has never been attacked by them except in one instance. This was in the autumn of 1842, as observed in a former part of this work,† when the grubs prevailed so as completely to destroy immense patches of the grass, the layer appearing as if burnt, and tearing up in large flakes with the slightest pull. In fact, the field was so denuded in different places, that at one time it was proposed to pare and burn the whole, and lay it down afresh.—This was not done, and it proved in the end unnecessary; for to our surprise the following year the grass was as green and flourishing as ever, and not a grub was to be seen: yet there was no reason to think these had been devoured by the rooks, which, as before stated, never found them

* *Naturalist's Calendar*, p. 100.   † See p. 148.

out; nor, on the supposition of their having worked deeper (being full-fed and ready to undergo their metamorphosis), was there any apparent increase of numbers of the perfect insect when midsummer arrived.

When flakes of grass are torn up beneath which these grubs are at work, the latter are found lying upon their sides, with the abdomen coiled upon itself; and it appears to be in this position that they feed, eating their way along from root to root.—In the above instance, they were observed continuing their havoc in this manner till the 18th of November; after which they retired from the surface, and went deeper into the ground, and were no longer seen.

These grubs (or one of a very similar kind) are not unfrequently destructive in the kitchen-garden, attacking particularly the young lettuces of autumn growth, which are intended to stand the winter. Wherever this is the case, the plants wither and die without any apparent cause, till they are pulled up; when they often separate just at the crown of the root, where the grub may be found actively at work, and which part it seems particularly to select.

GLOW-WORM.*

WHITE has observed† that glow-worms appear to " put out their lamps between eleven and twelve, and to shine no more for the rest of the night."— This is certainly not always the case; as some which

---

* *Lampyris noctiluca*, Linn.   † *Nat. Cal.* p. 118.

we narrowly watched in one instance, with a view to ascertain this point, continued their light till considerably after midnight.*

PTINUS FUR.†

THE destructive habits of this insect in museums are well-known. I have, however, had more reason, in my own case, to lament its ravages in my library; and I do not remember to have ever heard book collectors in general complain of its attacks in this way. A large number of my books, however, have suffered, principally in their bindings; and I do not find Russia leather any security, though its strong smell is thought by some to be prejudicial, or at least disagreeable, to insects in general. The leather of the bindings, however, is not the only part they devour: they often attack the paper and pasteboard of which the covers are made; and also the little cylindrical rolls, fastened down with thread, which appear at the top and bottom of most bound books.

* A more interesting fact, connected with the shining of these insects, has been lately made known, though it may require further observation to decide that this is generally the case; viz. that the female puts out her light immediately after her connexion with the other sex. It is well known that the female insect alone shines to any degree, and it has been supposed that she is endued with this property for the purpose of attracting the male, which is winged, she herself being apterous; and, if this is correct, it seems *à priori* probable that the light would be extinguished when its services were no longer required. See *Ann. des Sci. Nat.* 2nde série, tom. xviii. p. 379.

† *Stephens's Man.* p. 200.

In this last part particularly, the maggots may often be found secreted. I find no effectual means of checking this annoyance, except that of often taking the books down and thoroughly brushing them.*

DEATH-WATCH.†

I was once staying in an old house, where these insects prevailed to such a degree, that, during the spring, the walls resounded the whole day long with their continual rappings. It is generally supposed that the noise they make is intended as a call to the other sex; and it is curious to observe one of them labouring, as it were, to make itself heard. Raising itself on its hinder legs, it beats forcibly on the wall on which it stands with the fore part of the head, giving seven or eight strokes at a time in pretty quick succession. These are repeated at intervals; and, where the insects are numerous, after a while become irksome to the ear. The noise exactly resembles that made by gently tapping the finger-nail against the hard surface of a table; indeed, upon

---

* Sir Thomas Phillips, in a communication to the British Association in 1837, made mention of several insects that had attacked the books in his library, but these are stated to have been principally *Anobia*, which only devoured the paste used in binding, and which he recommended in consequence to be mixed up with a solution of corrosive sublimate. He makes no mention of the *Ptinus fur*, which, as above stated, attacks the binding itself, as well as other parts of the books. Corrosive sublimate can hardly be applied in this latter case. See *Report of Brit. Assoc.* vol. vi. (*Notices of Commun. to the Sections*) p. 99.

† *Anobinum tessellatum*, Fab. *Steph. Man.* p. 201.

doing this, a death-watch will frequently answer the call, if within hearing.* Where they beat long in one place, they make a brown spot to the size of a silver penny; and the paper of the room in which I resided was covered with such spots to a considerable extent.

Unless these insects are numerous in a house, it is not very easy to get a sight of them, since they generally keep behind the wainscoting or paper, and are, moreover, very similar in colour to old wood.— In this neighbourhood I have occasionally found them in decayed willows, but very rarely observed them in houses at all.

It is worth noticing that the *Anobium tessellatum*, which is the insect above spoken of, is not the only species of this genus that beats in the manner described. A similar noise, but much fainter, and not so readily distinguished, is made by the *A. striatum*,† the species which is so destructive to old furniture, and common in houses everywhere.

TURNIP WEEVIL.‡

IN the autumn of 1840, the turnips in this neighbourhood were greatly affected by the attacks of a small weevil, which causes the root to rise everywhere into knobs and excrescences. A farmer brought me several roots thus distorted, and in ap-

---

* This fact is mentioned by Kirby and Spence (vol. ii. 1st edit. p. 387); the above statement, however, rests on my own trial of the experiment. † *Steph. Man.* p. 201.

‡ *Ceutorhynchus sulcicollis*, Steph. Man. p. 224.

M

pearance blistered, not knowing the cause from which
it proceeded. Every one of these knobs, however,
contains a maggot, which is at once obvious on cutting into them. Being desirous of ascertaining the
species they belonged to, I put several of the diseased roots into a large glass jar, having some earth
at bottom, and covered with gauze at top to prevent
the escape of the perfect insects, whenever they
should appear. I waited three or four months in
expectation of their coming out, but none shewed
themselves. At the end of this time the turnips
were rotten and stinking; but, on examination,
the maggots were still in them and alive. At last,
quite at the end of the following June, when the
turnips were almost liquid from putrefaction, the
insects appeared in large quantities. They proved
to be the *Ceutorhynchus sulcicollis.**

LIXUS PRODUCTUS.†

THIS singular-looking and rather local insect is
very common in some parts of the fens of Cambridgeshire, inhabiting the stems of the great water-parsnep
(*Sium latifolium*), on the pith and internal membranes
of which the larvæ feed. During the first week in
August 1827, I noticed several of these last in plants

---

* They were so named by Mr. Stephens, to whom I shewed
them. Kirby and Spence, who have noticed this disease of the
turnip, (vol. i. 1st edit. p. 186,) attribute it to the *Rynchænus
assimilis*, F. (*Nedyus assimilis*, Steph.); perhaps there may be
more than one species of insect that causes it.

† *Steph. Man.* p. 252.

growing in the ditches between Ely and Littleport. When the larvæ are full-grown, they fall to the bottom of the cavity, the floor of which is formed by the joint of the stem next below them; and there, inclining themselves against the walls of their chamber in a nearly perpendicular direction, pass into the pupa state: the perfect insect, soon after its appearance, eats its way out of the plant. It is rare to find these insects in those plants which grow in the water, or at least to find them alive, as from the rising of that element in rainy seasons they are liable to be drowned; but they abound most in those ditches which are dry during the summer. These insects vary greatly in size, some being half as big again as others: the colour also is a little variable.

APION FLAVIPES.*

I OBSERVE annually that, about the month of September, several species of *Apion*, more particularly the *A. flavipes* and its allies, resort in great numbers to the evergreens in gardens, on which they may be found till the cold weather either kills them or compels them to hibernate. If the boughs be shaken over a white cloth, quantities will fall, though not perhaps previously noticed on the plants. In fact, they seem to confine themselves chiefly to the under surface of the leaves. For what purpose they frequent the shrubs, I am not aware; as the plants, in which these species of Apion are bred, and on which they feed in the larva state, are the Dutch and red

* *Steph. Man.* p. 259.

M 2

clover, belonging to quite a different natural family. The laurels appear, in general, to attract them most.

BRUCHUS GRANARIUS.[*]

Jan. 10, 1827.—A PERSON brought me to-day from a farm at Upwell, on the borders of Norfolk and Cambridgeshire, several specimens of the *Bruchus granarius*, which had been making great havoc among the beans in that neighbourhood. At what exact time the attack commenced, and how long the mischief had been going on, is uncertain, as the evil was not perceived till the farmers began thrashing out their beans, which had been stacked the preceding harvest. It then appeared that more than one-half of the crop had been entirely destroyed by these insects, which had eaten into the heart of the kernel.[†] Some of these beans I examined, and found to consist of little else than the mere outward husk, with a circular hole on one side, through which the insect, after having devoured the contents, had made its escape. In some cases the insects were still within, and were observed coiled up, immediately beneath the outer coat of the seed. I presume the mischief in this instance was principally effected by the grubs or larvæ of these insects, which had been bred in the beans from eggs originally deposited there by the parent; and this is according to what I find stated

[*] *Steph. Man.* p. 266.

[†] According to Kirby and Spence, (vol. i. 1st edit. p. 175,) the mischief caused by the Bruchus in this country is seldom very serious; it was decidedly so, however, in the above instance.

in books : yet it is worthy of observation, that all the
beans which came under my inspection, as well
those which had the perfect insect still in, as those
from which it had escaped, had the same round hole
opening externally. This looks as if in some cases
the beetles had entered by this aperture ; since, if they
had been bred in the kernel, the holes, one would
think, would not have been made till the time of
their leaving the beans altogether.*

It is to be regretted that the study of noxious
insects is not more attended to by practical agricul-
turists. It is often such persons alone who can supply the facts necessary for clearing up their history.
And the extent to which they suffer in their crops
from the attacks of different species in certain seasons one might have supposed a sufficient motive for
undertaking the inquiry. Something has been done
of late years in this way; but a vast deal more of
investigation is needed to put us in the way of suc-
cessfully counteracting these enemies, so as to prevent the immense damage they occasion. They may
appear puny and insignificant when viewed singly;
but, in their combined operations, they are often more

* The most recent notice of this insect I am aware of is by
Mr. Walton, in the *Annals and Magazine of Natural History*,
vol. xiii. p. 207. According to his statement, "the larva com-
pletes its metamorphosis within the seeds, consuming a considerable portion of the interior." It appears also that it sometimes
attacks peas. Mr. Marshall is said to have "observed in a barn
in Kent a quantity of peas infested with this beetle, which had
destroyed nearly half the crop ; in every pod that he opened he
found an insect, and the exterior part of the peas was more or less
consumed."

destructive and alarming than other animals infinitely superior to them in size, and ranking far
higher in the scale of nature.

HELODES PHELLANDRII.*

THIS insect is found within the hollow stems of
the water-hemlock (*Œnanthe phellandrium*), feeding
upon the inner coats of that plant, much in the same
way as the *Lixus productus* does on those of the
Sium latifolium. In August 1827, they occurred in
the greatest profusion in the fens about Ely. Every
plant of the water-hemlock that was to be found had
the stems crowded with them: vast quantities,
in some instances, were drowned by the rising of
the water above that joint of the stem in which
they were confined. As the *Œnanthe phellandrium*
almost always grows in the middle of the water,
these insects must be often subject to this accident.

COCCINELLA GLOBOSA.†

April 26, 1824.—I OBSERVE the *Coccinella globosa*
in great abundance just at this time on the young
plants of the field campion (*Silene inflata*) under the
hedges. They appear to be particularly attached to
them. This is in general one of the rarer species of *Coccinella* about here. When touched, they fall instantly
to the ground, and lie so concealed in the grass that
it is difficult to secure them. They vary greatly in

* *Steph. Man.* p. 311.　　　　　† *Id.* p. 317.

the number of the spots, which are also sometimes
so much larger than usual as to be nearly confluent.
A few specimens may be observed with the elytra of
a uniform dull red without any spots at all.

MELOE PROSCARABÆUS.*

In the spring of 1836, I found a large female of
this species on a bank, apparently in the act of depositing its eggs. On removing it to a glass jar, with
several inches of damp earth at bottom, it continued
laying its eggs; from which, in about a fortnight's
time, (as near as I remember, for I have no note
of the exact period,) were hatched a large number of
those small hexapod insects, similar to what are found
parasitical on bees, and concerning which entomologists have been so much puzzled, as to whether they
are the real larvæ of the meloe or not. These were
of a rufo-testaceous colour, and about three-fourths
of a line in length, or barely so much. They are
certainly identical with a parasite of exactly the
same form, size, and colour, which I have frequently
observed on bees on the Devil's Ditch, as well as
swarming on the blossoms of the dandelion, and
other plants growing in that locality. It is worth
noticing, however, that there appear to be more than
one species of this hexapod parasitical on hymenopterous insects. I have seemingly two distinct ones
in my collection, which occur equally often in this
neighbourhood : one of these is that just noticed as

* *Proscarabæus vulgaris*, Steph. Man. p. 335.

identical with what was bred from the eggs of the *Meloe proscarabæus;* the other is larger, of a much darker colour, approaching to black, though with the legs in part testaceous, and with the setæ at the apex of the abdomen longer. Perhaps this last is only the former in a more advanced state of growth. If, however, they be distinct, they must come from two different species of *Meloe,* supposing them really to be the larvæ of this genus of beetles. The only species of *Meloe* found about here, besides the *M. proscarabæus,* is the *M. violaceus;* and this is not at all abundant.*

EARWIGS.†

EARWIGS, as everybody knows, creep much into houses during the summer months, and often secrete themselves in the daytime in the crevices about windows. But in two or three instances I have observed them to be particularly numerous in such places after painting the outside of the house. I fancied the paint had some attraction for them, as almost before it was completely dry I noticed them crowding upon the sill of the window beneath the bottom of the sash, when the latter was down.

* For further information on this mysterious subject the reader is referred to De Geer, *Hist. des Ins.* tom. v. pp. 8—12. tab. i. fig. 5—8 ; Kirby, *Mon. Ap. Angl.* vol. ii. p. 168; and Westwood, *Entom. Trans.* vol. ii. p. 184.—See also *Entom. Trans.* vol. iii. p. 294, where (since writing the above) I find an allusion by Mr. Smith to two kinds of these supposed larvæ of the *Meloe ;* probably the same as those noticed by myself.

† *Forficula auricularia,* Linn.

GRASSHOPPERS.*

THE smaller species of *Locusta*, constituting the
grasshoppers of our fields and meadows, have not been
much attended to by entomologists, and their characters are but ill-defined. Many of them too are subject to much variation of colouring, so as to render the
distinguishing of species difficult. It has, however,
occurred to me, that their different notes might assist
much in the determination of this last point. As far
as I have observed, the same species always emit the
same notes. As insects in general are not in the
practice of emitting regular sounds, there are but
few groups in which this habit could be availed of
for such a purpose;—but in this family and some
other allied ones, it seems as if it might be made
serviceable.†

During hot summer weather, I observe that field
grasshoppers *crink* (to use White's expression) during
the greater part of the night.‡

* *Locustidæ*, Leach.

† Since this was written, I find a similar observation has been
made by Siebold, a Prussian entomologist, who has remarked,
with respect to the *Gomphoceri*, that "the individual species are
easily recognized by their chirping," each having its own peculiar
way of producing the noise. See *Reports on the Prog. of Zool.
and Bot.* published by the Ray Society, p. 218.

‡ The same is observed by White of the chirping of the field
cricket. *Nat. Hist.* Lett. XLVI. to D. Barrington.

DRAGON-FLIES.*

Aug. 22, 1825.—DURING a walk in the fens to-day, we observed vast quantities of female dragon-flies, of one of the larger kinds, that kept up an incessant hovering over the ditches, engaged in laying their eggs. The operation was rather curious. They poised themselves in the air for several seconds in the same spot, and then suddenly darted down towards the surface of the water, alighting generally upon the lower part of some reed, or on a piece of turf that lay just covered by the water, where they would remain a short time with their abdomen immersed quite up to the thorax. During this interval I conceive the egg was being deposited. The process was then repeated, and continued by the same insect as long as we had patience to remain.

MASON-WASP.†

July 14, 1823.—WE observed this morning one of the species of mason-wasps, as they are termed by entomologists, carry off a small caterpillar, and deposit it in a hole in a wall adjoining, where we found it had constructed a nest of mud and sand of the size of a small walnut. On breaking through the outer crust, we found several other caterpillars of

* *Libellulidæ*, Leach. The species above noticed was one of the larger kinds of *Æshna*.

† The insects noticed under this head are probably either of the genus *Odynerus*, Lat., or *Epipone*, Kirb.

a similar kind, which had been previously stored up as food for its future young, according to the well-known habits of this genus of insects. We partially destroyed the nest whilst examining it; but, to our surprise, on our return to the spot a few hours afterwards, we found it completely restored to its original state. We destroyed it again, and it was again rebuilt; but on pulling it to pieces a third time the insect flew out, and shortly afterwards returned accompanied by another of the same species (supposed to be the other sex), whom, as it were, it had gone to fetch in order to assist in the rescue, for they together seized up the caterpillars with their jaws and feet, and bearing them off as well as they could flew out of sight, never again to revisit the ill-fated spot. With these caterpillars no doubt their eggs would have been deposited; and the whole proceeding is strongly indicative of the affection and assiduity with which these insects will labour on behalf of their young.

There is a small green caterpillar, which, it is well known, is particularly selected by one species of solitary wasp, in order to be placed along with its egg, as food for the future young. But what very much strikes me is, that, at a certain period of the summer, I suppose when these wasps are engaged in nidification, the above caterpillars may often be observed in numbers scattered about the ledges of the windows in my house, and, as it would seem, deposited there without a purpose, since there is no appearance of a nest to be found near them. I believe these caterpillars to be brought by some species of *Odynerus,* as the circumstance generally occurs at the time that these insects are in the habit of entering houses;—but why they should be left there, and deserted, in such numbers, and not availed of for the purpose for which they were originally carried off, I am at a loss to conjecture.—Is it because they are discovered afterwards to be too small, and not full-fed,*—or from any circumstance unconnected with the particular state of the caterpillar ?—Be the cause what it may, there the caterpillars remain till they waste and die.

————

I conceive it to have been some species of mason-wasp, which was once the cause of much inconvenience to me, from the circumstance of its constructing its nest in the lock of a drawer. It was on the 26th of September 1837, that, on my return home, after an absence of six weeks, I attempted in vain to

* Kirby and Spence seem to speak as if the mason-wasps generally selected full-grown grubs as food for their future young. They observe, that " if those that are but partly grown were chosen, they would die in a short time for want of food, and putrefying would destroy the inclosed egg, or the young one which springs from it. But when larvæ of any kind have attained their full size, and are about to pass into the pupa state, they can exist for a long period without any further supply. By selecting these, therefore, and placing them uninjured in the hole, however long the interval before the egg hatches, the disclosed larva is sure of a sufficiency of fresh and wholesome nutriment." Vol. i. 1st edit. p. 341.

introduce a key into the lock of the table-drawer in my study, in order to open it. Finding there was an obstruction, I introduced a sharp pick, and, to my astonishment, drew out a number of small gravel stones and lumps of dirt, more than sufficient to fill a teaspoon. At first, it was supposed that these had been intentionally pushed in by some one mischievously disposed; and it was not till a carpenter had come and forced the drawer open, and taken the lock off (which was necessary in order thoroughly to clean it), that the real cause of the annoyance became apparent. It was then discovered that some insect had entered by the key-hole on different occasions, during the time, as was supposed, of the window of the room being open, and constructed its nest within the wards of the lock; a considerable portion of the nest being still entire, and in it a solitary larva, to all appearance nearly full-grown. Not more than one larva was observable; if there was a second, it must have been so small as to escape notice: neither was there any store of food found with it; but probably this, whatever it may have been originally, had already been devoured.

Since writing the above, Professor Henslow has sent me the particulars of a very similar case, that occurred not long since in his parish of Hitcham, in Suffolk. He says: "The padlock upon the door of our village coal-house was brought to me, in consequence of the key having been broken in an effort to overcome the resistance opposed by some sand, which it was believed at the time had been inserted by a mischievous boy. Upon opening the padlock, it was found the obstruction had been occasioned by a hymenopterous insect, apparently some species of bee, having selected this retreat for the construction of its nest. The sand was intermixed with a large quantity of pollen, which the bee had laid up in store ; but the persons who brought it to me had neglected to observe whether the larva was still in it." *

COMMON WASP.†

THE large female wasps which occasionally appear in the spring, in April and May, are well known to

* Several cases similar to the above are on record by different authors.—Thus, Colonel Sykes speaks of a species of *Sphex* in India, which makes its earthen nest within the locks of the doors, and blocks up the key-holes. *Reports of Brit. Assoc.* vol. vi. p. 252.—There is also mention made in *The Zoologist* (pp. 265-6), of a fossorial hymenopterous insect that deposited several green caterpillars in succession, to the number of half a dozen, between the open leaves of a book (each immured in a cell of moist clay), which chanced to be standing on one end near a window. The book was unfortunately disturbed afterwards, so that the caterpillars were not reared. This insect is thought by the editor to have been most probably the *Epipone lævipes.*—Mr. Curtis, however, mentions a case, which seems similar to this last, in which the insect was reared, and proved to be the *Odynerus parietinus.* He says, "My friend Mr. Charles· Fox detected upon the top of a book, across which another was laid, some cells of a somewhat triangular form, covered externally with mud, and formed of a silky substance within : he very obligingly transmitted the book to me last winter, and in the spring nearly twenty specimens of the insect figured (*Odynerus parietinus*) made their appearance ; they were all females, and did not vary in the least."—*Brit. Entom.* tab. 137.

† *Vespa vulgaris,* and *V. rufa,* Linn.

be the founders of the nests from whence come the wasps of the succeeding summer. It is, consequently, a great matter to destroy such as can be found at this season. — In general, however, we only meet with one here and there. I was therefore surprized by a relative of mine informing me, one year, that he could obtain twenty a-day by beating a whitethorn hedge near his house, on one side of which was a pasture and on the other a road. This was in the middle of May 1842; and many of the wasps so procured were submitted to my inspection, and proved to be the females of the *Vespa vulgaris* and the *V. rufa*, in about equal quantities. It was observed that few could be obtained anywhere else, and there seemed no peculiarity in the site or composition of the hedge to bring it into favour. Was this an accidental circumstance? or is it usual for these queens to secrete themselves in hedges at this season more than in other places?—If the latter, one would think they might be obtained without much difficulty and destroyed.* It may be noticed, that when the above wasps were taken, the weather was fine and seasonable, but without much rain, and inclining to drought. The preceding winter had

* Since writing the above, I find a note by Mr. Westwood, which seems to throw some light on this circumstance. He mentions his having "observed the excessive fondness of wasps for honey-dew upon whitethorns in the spring;" and he has suggested whether it might not be "advisable to watch situations in which *Aphides* abounded at that time, in order to destroy the queen-wasps attracted to such spots." See the Proceedings of the Entomological Society, as reported in *Ann. and Mag. of Nat. Hist.* vol. xvii. p. 66.

been a moderate one, neither particularly mild nor severe. The summer following was very fine (August especially very hot), and wasps were in great abundance.*

———

Queen-wasps sometimes appear abroad as early as the end of March. One that had hibernated among some loose papers in one of the rooms in my house, in which there had been no fire during the winter, first began to stir its legs and antennæ on the 26th of that month; the following day it changed its place, and on the 28th took wing, and was seen no more. This was in 1839. It would seem as if the period of their first flight was determined by something else besides temperature. The mean temperature of the above 28th of March was not more than 45° of Fahrenheit, and several days during the week previous had been quite as warm. It deserves mention, however, that the air was in a very electric state on that day, as indicated by frequent peals of thunder (the first that had been heard that year), attendant upon showers. The weather generally during March

* It does not always follow that a large number of queen-wasps in the spring portends an abundant supply of workers in the autumn. " It sometimes happens" (Kirby and Spence write) "that when a large number of female wasps have been observed in the spring, and an abundance of workers has in consequence been expected to make their attack upon us in the summer and autumn, but few have appeared. Mr. Knight observed this in 1806, and supposes it to be caused by a failure of males." Kirby and Spence think it may be sometimes owing to wet seasons, by which the nests become inundated. Vol. ii. p. 111.

1839 was cold and ungenial, and the spring of that year backward.

I observe that the sliding rods, upon which the summer blinds outside my window run up and down, are particularly resorted to by wasps in the month of July, and the wood very assiduously scraped off with their jaws in order to be employed in the fabrication of their nests. Both the common wasps (*Vespa vulgaris* and *V. rufa,*) and also the tree-wasp (*V. holsatica*) visit them for this purpose. These rods are of ash, and are quite scored in places by the marks left by their jaws.—It should be mentioned, however, that they have been up several years, and their surface has been a good deal roughened, and acted upon by the weather.*

---

* Any little facts of this kind appear worth preserving, as, according to entomologists, there has been some difference of opinion respecting the kind of material employed by the common wasp in the construction of its nest. See Mr. Newport's remarks on this subject in the *Entomological Transactions* (vol. iii. p. 190). He says, "Reaumur states that the wasp procures its material from decayed timber, like the hornet; but White of Selborne, and Kirby and Spence, assert that hornets alone obtain it from rotten or decayed wood, while the wasp procures it from *sound* timber. From my own observations I can state most positively that the wasp procures at least some portion of the materials it employs from *rotten* wood, as I have many times witnessed during the last summer. I saw both the common wasps and the hornet busily engaged at the same moment in obtaining materials from the same piece of rotten wood. The wasps even penetrated into the soft wood in several places to procure the material. But I have also

The following notes respecting wasps and their nests were sent me, at my request, by my brother-in-law Professor Henslow, who has paid a good deal of attention to these insects.

He says, in a letter, dated Hitcham, December 1845: " I have occasionally amused myself with taking the nests of wasps and bees, and have been interested with their economy, though I fear I have not much to state that can be called positively new to you. The soil here is a very stiff clay, well filled with small rolled lumps of hard chalk and stones. The wasps, in carrying off the earth to enlarge the holes in which their nests are suspended, cannot remove these lumps of chalk and stones; which, by their working round them, become detached, and keep continually subsiding to the bottom, as the cavity is enlarged, until at length they form a sort of loose pavement. The droppings which fall from the nest upon this pavement are sought after by flies, and a number of their larvæ may generally be found crawling among the loose stones. Sometimes a large stone projects from the side of the hole, and becomes gradually inclosed as the nest enlarges, until at length it has been completely detached from the soil, and is then found suspended among the comb in the very middle of the nest.

seen the wasps, as many others have done, procuring it from the solid wood of a window framing; although, it must be remarked that the wood in this instance also has been that which was some-what affected by the weather."—See more on this subject in the communication which follows, made to me by Professor Henslow.

" *Vespa rufa* appears to be nearly, if not quite, as common here as *V. vulgaris.* I have found large nests of each within a few feet of each other, and each kind retaining throughout its structure its characteristic peculiarities; that of *V. vulgaris* being wholly composed of materials prepared from sound fibre, and that of *V. rufa* manufactured of much more tender material, like that of the hornet (*V. crabro*), and which is believed to be from rotten wood: though I think it may be worthy of closer inquiry whether it be not really obtained from the *bark* of trees. I possess an example of *V. vulgaris* having prepared a portion of the outer case of its nest from a piece of blue paper. The paper had been masticated and worked up exactly as the fibre, but has retained its colour. There is a marked difference between the appearance of the larvæ of the *V. vulgaris* and the *V. rufa.* The latter are (proportionably with their imago) larger, plumper, and whiter. It is sometimes difficult, if not impossible, to distinguish *V. rufa* by the anchor-shaped mark on the face, which is sometimes nearly obliterated."

TREE-WASP.[*]

On felling a fir-tree, last autumn (1845), in one of the plantations at Bottisham Hall, a nest of the tree-wasp was found suspended from one of the twigs, about halfway from the ground. The lower-most portion, containing the entrance to the nest,

[*] *Vespa holsatica,* Linn. ? or *V. britannica,* Leach ?

appears to have been broken off in the fall : when complete, the whole must have been nearly the size of a child's head. What remains is of a somewhat hemispheric form, approaching to pear-shaped at its upper extremity where united to the twigs ; it is about six inches in diameter, and nearly eighteen in circumference. The outer case is formed of several layers of a papyraceous substance, easily separable one from another, and, on the whole, very similar to the external covering of the nest of the common wasp. On examining this substance under a high lens, it appears to be made up of minute filaments irregularly matted together, and afterwards besmeared with some viscous fluid causing them to adhere. Within the nest were four tiers of cells one above another : the lowermost tier alone, which was much the smallest, consisting of upwards of one hundred and thirty. Some of these cells are much bigger than others. Many of them were open, and contained larvæ ; others had the mouths sealed up, and contained pupæ, or the perfect insect just ready for exclusion.

It seems remarkable that, at this advanced period of the year, so large a portion of the brood should be still in the immature state. One can hardly imagine that the whole would have completed their metamorphosis before winter, had the nest remained undisturbed. Probably, however, nidification continues to go on till the cold puts a check to further operations, when those larvæ, which are the latest in hatching, must necessarily perish.

HORNET.*

A FRIEND of mine gave me, some time back, a small nest, supposed to be that of the hornet in an incipient stage, which was found suspended, in the month of June, from the top of the inside of an empty beehive, at Mickleham in Surrey. In shape it is very similar to that of the *Vespa holsatica* last described, but much smaller, and very differently constructed. Its size is about that of a large orange. The outer casing consists but of a single layer, which is thin and porous, the colour brown, yellowish-brown towards the top, and similar to that of decayed wood. This casing, when closely examined, is found to be made up, not of filaments, but of minute abraded chips and fragments of rotten wood, opake, and apparently very little altered by any kneading process. There are not more than five-and-forty cells in all, forming a single layer at the upper part of the nest, the walls of which extend far below it, as if to allow of other layers being added. These cells are very much larger than those of the tree-wasp, but not so firm in texture. When the nest came into my hands, they were all empty. This appears to have been the first commencement of a nest by the mother-queen alone, previously to having any workers to assist her in the undertaking.

* *Vespa crabro*, Linn.

BEES.

Sept. 13, 1825. — I HAD often remarked that the blossoms of the dahlias in gardens contained bees apparently dead, but had not reflected much on the circumstance, till I lately heard the same observation made by another person. I have since paid more attention to this matter, and have watched bees entering the flowers of these plants, for the purpose of sucking the nectarium, or collecting the pollen, when they were obviously soon seized with a sort of torpor; in which state, if not speedily removed, they often died. This appears to result from some poisonous quality in these flowers detrimental to bees : they should not therefore be planted near their hives. When these plants are in full blossom, numbers of bees may be found in this state of intoxication. Humble-bees, however, seem to be more influenced by its narcotic powers than hive-bees.*

In the early part of the spring, when there is no great abundance of flowers, hive-bees are very much

* The above observation, made by myself many years back, has been since made by others, as appears by the following extract from one of the numbers of the *Gardener's Chronicle* :—

"BEES AND DAHLIAS.—The cultivation of the dahlia is incompatible with the success of the bee-keeper. For many years I was very successful with my bees, having upwards of twenty hives yearly, and of course abundance of honey ; but from the time that I commenced growing dahlias the bees declined, and I had at last to give up those useful insects altogether. They became intoxicated by feeding on that flower ; many of them I found

attracted by the box, and resort to it in great numbers as soon as it is in blossom. I was struck, however, by the fact that, in 1844, they seemed impati-
ent of waiting till the buds had expanded, and actually proceeded with their jaws to force them open,
to get at the immature pollen.

I observed very many engaged in this process,
wherever there were box-trees, on the 29th of
March, and for several days afterwards.

BUTTERFLIES.

Papilio machaon.*—This splendid butterfly was
formerly found in the greatest profusion throughout
the whole extent of the fens between Cambridge and
Ely. It is still in tolerable plenty some seasons ; but
the numbers appear annually diminishing, from its
being so much sought after. Large numbers are
caught by the poor people, in the hope of disposing
of them to collectors. It is probable that in a few
more years it will become rare. There appear to be
two broods in the year; the first shewing itself on
wing from about the middle of May to the end of
June, the second from the middle or end of July
to the middle of August.† The larvæ are found

dead in the blossoms, or lying on the ground underneath, and
those which got home formed little or no honey. I have heard
the same remark made by many persons to whom I mentioned it,
both in England and Ireland."—*Notice by a Correspondent.*

* *Steph. Illust.* (*Haust.*) vol. i. p. 6.

† Stephens is rather inclined to the supposition that there is a

principally on the marsh milk-parsley (*Selinum pa-lustre*), which is undoubtedly the plant to which they are most attached, though in confinement they will feed readily on several species of umbellatæ.

Gonepteryx rhamni.†—Some plants of the red valerian, which grow in my garden, are much visited by brimstone butterflies in the autumn. On a fine bright day in September, they are often covered with these insects, which seem to prefer them to any other. In what the attraction consists I am not aware.

Pontia cardamines.‡—The disproportion of sexes in the orange-tip butterfly is very remarkable. The males may generally be observed in plenty in lanes and meadows, making their appearance with great regularity the first or second week in May; but the females I have never taken, at least in this neigh-bourhood, more than a very few times : perhaps it is owing to the latter being more inactive, and less fre-quently on wing.

Vanessa urticæ.§—The tortoise-shell butterfly, and the peacock (*V. io*)‖, though they hibernate like the rest of the *vanessæ*, are sometimes on wing in the open air on fine days in winter.—In December 1843, when the weather was remarkably fine and mild for the season, I observed both those species together flying briskly in the park at Bottisham Hall. Only one or two nights' frost occurred during that month :

constant succession of individuals during the above period rather than two distinct broods.

† *Steph. Illust.* vol. i. p. 8.   ‡ *Id.* vol. i. p. 23.
§ *Id.* p. 43.   ‖ *Id.* p. 44.

the wind was generally W. or S.W., and the thermometer occasionally as high as 54°.*

Vanessa antiopa.†—This insect, as is well known, is remarkable for the irregularity of its appearance in this country. I have a manuscript note by a naturalist, a relation of mine, now dead, in which it is stated that in August, 1789, this species appeared in the greatest profusion throughout Norfolk and Suffolk: great numbers are said to have been found drowned on the coast, and it was the general belief at the time that they had come over from the continent.

Of late years, a few specimens of this beautiful *Papilio* have been taken in this neighbourhood: one shewn to me in the autumn of 1842 had been captured a short time previous at Great Wilbraham, within three or four miles of Swaffham Bulbeck.

Vanessa atalanta.‡—Admiral butterflies are much attracted in the autumn by ripe fruit, and especially by mulberries.§ One of these trees, which formerly grew near here, was generally visited daily, in the month of September, towards noon, when the sun

* I find a notice in *The Zoologist* (p. 64), of the peacock butterfly being seen on wing at Lavenham in Suffolk, Dec. 13, 1842: also (p. 113), of the little tortoise-shell flying about briskly at Dover, Dec. 15, in the same year.

† *Steph. Illust.* (*Haust.*) p. 45.

‡ *Id.* p. 46.

§ I find the fact of admiral butterflies feeding upon the juices of the autumnal fruits mentioned by Mr. Knapp, in his *Journal of a Naturalist* (3rd edit.), p. 289.

was warm, by swarms of these insects, whose rich tints of red and velvet black had a fine effect when presented to the light. I never particularly noticed any other species of butterfly about this fruit.

DEATH'S-HEAD HAWKMOTH.*

Sept. 23rd, 1830.—A FULL-GROWN caterpillar of this immense moth was brought me to-day from a potatoe-garden in this village. We had the curiosity to weigh it, and found its weight no less than four drachms and forty-two grains.

These caterpillars have become not unfrequent of late years in this neighbourhood, from the extent to which potatoes are now cultivated;—but I never knew but one or two instances of the occurrence of the perfect insect.

VAPOURER MOTH.†

Oct. 2nd, 1839.—THE faculty by which the sexes find out each other among the insect tribes is very mysterious. Having bred a female vapourer moth from a caterpillar found in the park at Bottisham Hall, I put it alive into a small chip box covered with gauze, and set it in a room in my own house with the window open; and, in a few hours, two males had flown in at the window in search of it,

* Steph. *Illust.* (*Haust.*) p. 114. (*Acherontia atropos.*)
† *Id. Orgyia antiqua*, vol. ii. p. 61.

and alighted on the wall near where the box was placed.*

July 11*th,* 1826.—THE caterpillars of the pink underwing abound this year to an immense degree on the common ragwort (*Senecio jacobæa*), reducing the plants to mere skeletons by their ravages. The caterpillars of the mullein moth ‡ are almost equally plentiful on the plants of the common *Verbascum.* Both these insects, as far as I have observed, are periodical in their appearance, shewing themselves in vast quantities after an interval of a few years.

I have a singular variety of the *Callimorpha jacobæa* in my cabinet, taken near Cambridge, in which the bright red of the under wings is exchanged for a pale orange yellow.

HERALD MOTH. §

THIS insect, as is well known, makes its appearance generally in the autumn, and hibernates during

---

* This faculty, possessed by some of the *Lepidoptera* to a great extent, is well known to collectors, who often avail themselves of this way of obtaining the male sex of certain species. See some remarks by Haworth on this practice, with amusing details, in his *Lepidoptera Britannica,* p. 82.

† *Callimorpha jacobæa,* Steph. *Illust.* (*Haust.*) vol. ii. p. 90.

‡ *Cucullia verbasci,* Steph. *Id.* vol. iii. p. 85.

§ *Calyptra libatrix,* Steph. *Id.* vol. iii. p. 50.

winter. Its torpor and listlessness, however, in
some cases, before the numbing effect of the winter's
cold can be possibly felt, as well as after it is over,
are very striking. I was much interested, in the
autumn of 1842, with a moth of this species, in
which this sluggishness prevailed to an unusual
degree. I first observed it sticking to the inside
wall of an outbuilding about the middle of September, evidently just emerged from the pupa state.
The weather was at the time very warm and seasonable, and the temperature of two or three days
consecutively as high as 70°. It was thought, of
course, that it would only rest there during the day,
and take wing in the evening. Yet, to my surprise,
it remained day after day in exactly the same position as when first noticed. This induced me to
watch it more narrowly, which I continued doing
throughout the whole winter, and a part of the following spring; and during all this time it never
changed its place once, or even shewed signs of ani-
mation, except when gently touched, when it would
move its head and antennæ a little; but, on the
finger being withdrawn, would quickly relapse into
its former state of repose. It was not till the 20th
of April, 1843, that it began to move. I then observed that it had shifted its quarters to another part
of the wall: the following day it had flown to the
ceiling of the building, where it remained over the
22nd; and on the 23rd it was gone. This last day
was the hottest we had had that spring, and the
temperature rose to 68°. Thus this moth was in a
half-torpid state over a period of seven months,

during which time the weather was often extremely mild, and December and January, considered as winter months, quite unseasonably so.—One is almost irresistibly led to ask in what state must be the volition and feelings of an animal, which, under such circumstances, for more than half a year, is not roused from its slumbers by any of the usual calls to locomotion and activity.*

GNATS (CULICES).

THE gnats which bite, and which are so trouble-some in houses, consisting of the *Culex pipiens* and two or three allied species, appear to me to be almost all females: it is rare to be bitten by the males at all.† These last keep more out of doors; and I observe, that the broods which may be seen

* The inactive habits of this insect, and its seldom taking wing, may be the cause of its being found in such perfect condi-tion so many months after its first appearance in the moth state. Thus, a writer in *The Zoologist* (p. 260) mentions taking one in an empty house on the 13th of March ; and observes that "it was thoroughly perfect, the scales not being abraded on any part of it." He seems in doubt whether it had hibernated or not: this however had most probably been the case. See also p. 333 of the same work, for a further notice respecting the hibernation of this insect.

† This fact is noticed by Humboldt with respect to the *Culices* of South America. He says, "the males are extremely rare, and you are seldom stung except by females."—*Pers. Narr.* vol. v. p. 98.

Humboldt mentions also another fact, with respect to one spe-cies, called in South America the *Zancudo*, which, if generally true of this tribe of insects, is curious, and interesting to know ;

playing about in the air on a summer's evening consist exclusively of this last sex.

Oct. 5, 1828.—Stinging gnats of various species,
but more particularly the *Culex pipiens* and the
C. annulatus of Meigen, abound in houses this autumn to a degree far surpassing anything I remember in former years. The cellars especially
are swarming with myriads, which blacken the walls
and ceilings with their numbers. This I attribute
to the wet summer.*

and that is, that if, when it has once fixed its sucker, " it be left
to suck to satiety undisturbed, no swelling takes place, and no
pain is left behind." He says, " We often repeated this experiment on ourselves in the valley of the Rio Magdalena, by the advice of the natives. It may be asked, whether the insect deposit
the stimulating liquid only at the moment of its flight, when it is
driven away; or repump the liquid, when it is left to suck as much
as it will. I incline to this latter opinion; for, on presenting
quietly the back of the hand to the *Culex cyanopterus,* I observed
that the pain, very violent in the beginning, diminishes in proportion as the insect continues to suck; and ceases altogether, when
it voluntarily flies away."—*Id* p. 114.

I have not noticed anything like this in the case of our English
Culices. Neither did I find, on mentioning Humboldt's statement
to Sir Robert Schomburgh, who has travelled over the same parts
of South America, and felt all the annoyance experienced by
Humboldt and Bonpland from the attacks of these insects, that
he could confirm it by his own observation.

* However gnats may occasionally abound, and annoy us, in this
country, we can form but a small idea of their immense numbers
in some other countries, particularly Equinoctial America. Humboldt observes, that " persons who have not navigated the great
rivers in that part of the world, for instance, the Oroonoko and
the Rio Magdalena, can scarcely conceive, how without interrup-

GNATS (CHIRONOMI).

WHITE has noticed the immense swarms of gnats
which are sometimes observable in the fens of the
Isle of Ely, and their resemblance to smoke.*   In
the autumn of 1843, such a cloud of these minute
insects was seen rising from the top of the west
tower of Ely cathedral, and the deception was so
great, that an alarm was raised, under the idea that
the cathedral was on fire.   It was not till persons
had ascended to the top of the building that the
cause of the appearance was satisfactorily ascertained.
I did not see any of these insects myself, but I was
informed by a gentleman of that place that some

tion, at every instant of life, you may be tormented by insects
flying in the air, and how the multitude of these little animals
may render vast regions almost uninhabitable.   However accustomed you may be to endure pain without complaint, however
lively an interest you may take in the objects of your researches, it
is impossible not to be constantly disturbed by the moschettoes,
zancudoes, &c., that cover the face and hands, pierce the clothes
with their long sucker in the form of a needle, and, getting into
the mouth and nostrils, set you coughing and sneezing whenever you attempt to speak in the open air."—*Pers. Narr.* vol. v.
p. 87.

The above illustrious traveller has dwelt at considerable length
on the subject of the moschettoes in his work, to which we would
refer the reader for the sake of the interesting details he has
brought together relating to it.   He speaks of the importance of
this subject, from the great influence which these insects exert "on
the welfare of the inhabitants, the salubrity of the climate, and
the establishment of new colonies on the rivers of Equinoctial
America."

* *Naturalist's Calendar*, p. 116.

had been shewn to Mr. Newman, the editor of the Zoologist, who pronounced them to be the *Chironomus prasinus*.*

———

Broods of a small jet-black species of *Chironomus* † may sometimes be observed in winter actually hovering over snow. From the freedom with which black surfaces radiate out caloric, one would think the colour of these insects must have the effect of reducing the temperature of their bodies, even below that of the air. Yet they seem almost as active, and as much at their ease, as if it were summer.

TRICHOCERA HIEMALIS.

THE power which some of the *Tipulidæ* have of resisting cold is very remarkable. During the severe winter of 1829-30, in the month of January, when the *mean* temperature of the twenty-four hours was varying from 28° to 33½°, and in two instances descended as low as 25° and 16° respectively, a brood of the *Trichocera hiemalis*, Meig., suddenly made their appearance in considerable numbers, settling upon the walls of different outbuildings, as if they had just emerged from the pupa state; and though they did not offer to take wing for several weeks,

———

* See a notice of similar clouds of dipterous insects being observed in Ireland, in *Ann. and Mag. of Nat. Hist.* vol. x. p. 6.

† Either the *C. byssinus* of Meigen, or some closely allied species.

yet they readily moved their quarters when disturbed; a proof that, notwithstanding the continued frost, they were not actually torpid.

FLIES.

BESIDES the true horse-flies,* which settle on horses in order to suck their blood, and which for this purpose have their mouths furnished with a most formidable set of knives and lancets,—there are hosts of others, which hover about them in swarms in hot weather, but which, having only a soft proboscis, without any instruments for cutting or piercing the hide, appear quite incapable of inflicting any sensible injury. These consist of various species of *Musca* and *Anthomyia* (Meig.), and perhaps other genera; and their sole object, as it appears to me, is to suck the perspiration of the animal when heated. They tease by their numbers, and by the constant titillation they keep up in settling; some horses, however, get habituated to them, and almost disregard them entirely.

———

If the leaves of different trees, but especially the lime and ivy, be examined in an autumn morning after a cold night, they will be found in many places covered with flies, having all the appearance of life, but which upon being touched are found to

* Different species of *Tabanus*, *Hæmatopota*, and *Chrysops*, Meig.

be dead and stiff, and what is more, in some cases
firmly glued down to the surface of the leaf by a
sort of cottony mildew. This seems owing to the
chill and dampness of an autumnal night coming
suddenly on, and surprizing such individuals as had
been tempted out of their hybernacula by the morning's sun; rendering them powerless, and unable to
effect their retreat in time. The same circumstances
promote the growth of the mildew, by which their
death is accelerated, if it be not mainly owing, in
the first instance, to the attacks of that fungus.
I observe this frequently about the latter end of
October or beginning of November, when the temperature of the air at sunset, especially if the sky be
clear, falls very rapidly, and radiation also exerts a
powerful influence in cooling down objects that are
much exposed. Sometimes flies may be observed on
windows, killed in the same way, and presenting a
similar appearance. In both cases, many of the individuals so nearly resemble living insects as almost
to deceive an entomologist; and I have occasionally
approached with unnecessary caution to entrap with
the forceps what I conceived to be a fly basking on
a leaf. *

* The above is extracted from a journal-book of mine bearing
the date of 1824. Since then the observation recorded has been
often made by others, and many suggestions offered in explanation of the fact. See *Loudon's Magazine of Natural History*,
(vol. vii. p. 530,) where there are several similar remarks by different observers. The cottony mildew above spoken of appears to
be a peculiar fungus found on flies only. Mr. Berkeley (in a letter
dated Dec. 5, 1837) informed me that he believed it to be the

Extraordinary swarm of flies.—During the month of September, in the year 1831, a small dipterous insect, belonging to Meigen's genus *Chlorops,* and nearly allied to, if not identical with, his *C. læta,* appeared suddenly in such immense quantities in one of the upper rooms of the Provost's Lodge, in King's College, Cambridge, as almost to exceed belief. The same species of fly, or one closely approaching to it, is not uncommon in most houses towards the decline of the summer; but in this instance their numbers were so great, and their appearance so sudden, as to surpass anything of the kind I had

Sporendonema muscæ of Fries ;* adding, that there was little doubt that the fly is attacked by it while yet living, but that the parasite is not fully developed till after death. The disease, it appears, may occur at any time, where circumstances are favourable for the growth of the fungus. The reason why it prevails most in the autumn is the dampness of the air at that season, which promotes the growth of all kinds of mould; and the suddenness with which the flies appear to be attacked is probably due to the rapid development of the fungus during the chill of an autumnal night. In the case of flies which are found occasionally, at all seasons, sticking to panes of glass, especially of windows that are not often opened to keep up a free circulation of air, the cause of death is the same, but it seems to come on more gradually.

It was probably individuals labouring under this disease which White saw, when (speaking of flies in houses in the decline of the year) he says, " at first they are very brisk and alert, but, as they grow more torpid, one cannot help observing that they move with difficulty, and are scarce able to lift their lges, which seem as if glued to the glass ; and by degrees *many do actually stick on till they die in the place.*"—*Nat. Cal.* p. 114.

---

* See *Eng. Flor.* vol. v. pt. ii. p. 350.

ever before witnessed. It was not till after a fortnight had elapsed from the time of these insects
being first noticed that I had an opportunity of
seeing them myself, during which interval their
numbers had been greatly thinned by fumigations
of tobacco and other substances employed as a
means of destroying them; nevertheless they were
still in immense profusion, and my informant told
me that in the first instance the greater part of
the ceiling, towards the window of the room, was
so thickly covered as not to be visible. The exact day of the month on which these insects first
shewed themselves was not noticed, but, as far as
could be remembered, it was about the 17th of
September. They appear to have entered the room
very early in the morning, by a window looking
due north, which had been open during a part
of the night, being first observed between eight
and nine A. M. A few were noticed in the adjacent rooms facing the same way, although, comparatively speaking, in no great quantity; perhaps,
in consequence of the windows of those rooms not
being opened at quite so early an hour. None
at all, however, had been seen in the house previously to that day. We are at present so ignorant of the habits and economy of the minuter
tribes of insects, that it is not easy to give an explanation of this phenomenon. It would be interesting to know whether the above had been all
bred in the immediate neighbourhood, and at the
same time, or whether they were swarms that had
collected from different quarters for the purpose of

migration. Many facts are on record which seem
to confirm the idea that insects do occasionally
change their quarters in immense bodies; and some
have occurred to myself, which, I have no doubt,
were connected with such a circumstance, not only
from the large numbers of the insects observed,
but from the steadiness of their flight, and their
continually persevering in one given direction. It
is worth noticing, with respect to the present
case, that King's Lodge is situate close to the
river Cam, which at that place runs nearly due
north and south; and it is just possible that this
circumstance may have had some influence in directing the movements of these insects. I find
also, by referring to a journal of the weather,
kept in the neighbourhood of Cambridge, that,
about the time when they were first observed, the
wind was N. N. W, and that it had been blowing
steadily from that quarter for four successive
days.*

ERISTALIS TENAX.†

FROM the middle of September onwards during
the autumn, these flies are much in the habit of
entering houses, where they prove tiresome in rooms
from the incessant humming they keep up. This
is so like that of a bee, and the fly itself so much

* The above account was published at the time in *Loudon's
Mag. of Nat. Hist.* (vol. v. p. 302), and is reprinted here.
† *Musca tenax*, Linn.

resembles a bee in general appearance, that persons who are not entomologists constantly mistake them for such, and call them drones. I observe that almost all the individuals that come into houses in this manner are, as in the case of gnats already alluded to, females.

ANTHOMYIA CANICULARIS.

IN the spring of 1837, a physician at Cambridge acquainted me with a case which had occurred to him a short time previous in his practice in that neighbourhood, in which large quantities of the larvæ of some dipterous insect were expelled from the human intestines. Some of these were submitted to my inspection, and proved to be the larvæ of the *Anthomyia canicularis* of Meigen. The patient was a clergyman, about seventy years of age. The symptoms of which he complained, previously to the first appearance of these larvæ, were—general weakness, loss of appetite, and a disagreeable sensation about the epigastrium, which he described as a tremulous motion. These symptoms commenced in the spring of 1836, and it was not till the summer and autumn of that year that the larvæ were observed in the motions. They then passed off in very large quantities on different occasions, the discharge continuing at intervals for several months. According to the patient's own statement, the chamber-vessel was sometimes half-full of these animals; at other times they were mixed with the stools. He thinks that altogether the quantity evacuated must have amounted to several quarts. The larvæ were all nearly of equal size, and, when first passed, quite alive, moving with great activity. The patient was not aware of having voided anything of the kind before. After the discharge ceased, his health improved, but was by no means perfectly re-established; and for some months after he was impressed with the belief that more larvæ were still in the stomach and intestines, though I never was informed in what way, or at what exact period, the complaint terminated.

It would be a matter of great interest, as well as importance, to ascertain the means by which these larvæ were introduced into the human body; but it is difficult to throw much light on this inquiry. It is observable that the symptoms of which the patient complained first shewed themselves in the spring of the year, which is the season in which, under ordinary circumstances, the larvæ would be hatched. The larvæ were not voided till the summer and autumn following, when they appear to have been nearly, if not quite, full-grown. Hence it would seem probable that they were conveyed into the stomach in the egg state, and that, after being hatched, they passed thence into the intestines; where they would have no difficulty in finding subsistence, if, as De Geer states of this species, they reside naturally, during this period of their existence, in the ordure of privies. But the question still remains, how, in the first instance, the eggs were introduced into the stomach; and I can only conjecture that they may have been deposited by the parent fly in some article of food, which had possibly become tainted, or was in an incipient state of decomposition, previously to being eaten.*

———

May 29, 1845.—I observed a species of *Anthomyia* this morning on the window of my study affected in a peculiar way, and similar to what I have before occasionally noticed in the common house-fly. It attracted my attention, from having remained a long time stationary without moving. On examining it closely under a lens of low power, it appeared to be engaged in slowly vomiting up a clear but slightly adhesive liquid from the extremity of the proboscis, and then as slowly drawing it back again into the mouth. This was repeated a vast number of times at short intervals. The globule of liquid, which was about the size of a large pin's head, remained at the extremity of the proboscis from one to two minutes, partly resting upon the glass, previously to being sucked up again. Just before its reappearance each time, the insect was much agitated, a violent tremor seizing its legs and *halteres* (especially the latter), though the wings remained motionless. Occasionally a drop of the same clear liquid was expelled from behind also, and at the same moment, and left upon the window-pane. The fly remained

* The above is abridged from a notice of this case, which was communicated at the time to the Entomological Society, and afterwards published in their Transactions, (*vol. ii. p.* 152,) where there will be found a full description, as well as figure, of the larvæ, together with references to one or two analogous cases previously on record.

half an hour engaged in this operation, and then flew away. During the time, it appeared to take but little notice of objects, and suffered me to place my finger very near it without moving. Whether the whole was the result of any accidental derangement of the system, or whether the regurgitation of the liquid in question was attended, as in the case of rumination in quadrupeds, by any pleasurable sensations, and effected at will, I am unable to determine.*

DUNG-FLIES.†

THESE flies, though bred in dung, are in their perfect state truly carnivorous, and devour other flies with greediness. Among the smaller species of *Anthomyia* they make great havoc, seizing them with their fore-feet, and then piercing their head with their proboscis to suck its juices.

LIVIA JUNCORUM.

THE alteration of structure induced in certain plants by the attacks of insects, so complete, in some instances, as almost to deceive at first the most experienced naturalist, is extremely curious. Nothing is more common than to find, during the latter part of the summer, in our fens, what at first view appears

---

* Since the above was written, I find a somewhat similar observation recorded in *Loudon's Mag. of Nat. Hist.*, vol. ix. p. 108.

† The flies here alluded to belong to the genus *Scatophaga* of Meigen.

to be a dwarfed variety of the common jointed rush (*Juncus articulatus*), but which is in fact the *nidus* of a peculiar insect, known to entomologists by the name of *Livia juncorum*. The larvæ of this insect, when hatched, feed upon the juices of the plant, the supply of which is increased by a greater determination of sap to the particular part punctured by the parent in laying its eggs; and the plant in consequence becomes monstrous at that part, producing compact leafy bunches, instead of stalks and flowers. If these be opened, they will be found full of the larvæ and pupæ; and by gathering them, and keeping them in a box for a few days, the perfect insect may be reared without difficulty. The latter appear about the second or third week in September.*

The galls, which occur so abundantly on oaks in the spring, as well as on other plants, are familiar instances of an analogous diseased growth of particular parts from the attacks of insects. Tubercles of a peculiar form may also be observed occasionally on the Scotch and spruce firs, very much resembling young cones, but which on being cut open are found to be the nest of a species of *Aphis*.†

* The first notice of this insect, so far as I am aware, by any English naturalist, will be found in the *Linnean Transactions*, (vol. ii. p. 354,) where there is also mention of the peculiar monstrosity which it occasions in the *Juncus articulatus*, and which was actually considered as a viviparous variety of that plant by Linnæus. Since then, both the insect and its nidus have been figured by Mr. Curtis in his *British Entomology*, (pl. 492,) to which work I refer those who desire further information respecting them.

† For notices of many other gall-like excrescences on plants

APHIDES.

Oct. 3rd, 1822.—THIS morning on rising we found
the air completely choked with *Aphides*. The steps
at the house-door, and even the very walls, were
black with them. On walking out, myriads alighted
upon one's clothes; and getting also into one's eyes,
ears, and nose, proved an intolerable nuisance. In
the middle of the day I took a circuit of about three
or four miles from home, but found the quantities of
these insects the same wherever I went. A friend,
too, who arrived from Cambridge, distant about
eight miles off, assured us that they were in equal
plenty there. Where could these prodigious multitudes come from, and whither were they directing
their flight? Such questions are easier asked than
answered. It is worth noting that the day was particularly mild and calm for the time of year, and had
begun with a fast mizzling rain, which lasted for a
considerable part of the morning. At 4 P.M. the
thermometer was as high as 64°. The wind was
easterly, and had blown steadily from that quarter
for three or four days previous.*

caused by insects, see Kirby and Spence (*Entom.* vol. i. pp.
448, 9.)

* A similar instance of large swarms of *Aphides* in a state of
migration is recorded by White, (*Nat. Hist.* Lett. LIII. to D.
Barr.,) as having occurred at Selborne, Aug. 1, 1785. It is deserving notice, also, that he particularly mentions that the wind
was then "all the day in the *easterly* quarter."

LETTUCE BLIGHT.

In the summer of 1844, the entire crop of lettuces in my garden were destroyed by a blight at the roots, arising from the attacks of a small species of *Eriosoma*.* This insect had never shewn itself there in any previous year to my knowledge. In this instance, all the young lettuces, from six to nine inches high, were observed with their lower leaves flaccid, and flat on the earth, as if parched from drought: the older ones, which had been tied up for blanching, were—some of them completely dead, and brown at the heart—others dying. No insects were observed upon the plants above ground; but, on pulling them up, the fibres of the roots were found thickly matted with a glutinous cottony substance, amongst which were crawling hundreds of the larvæ and pupæ. This was on the 28th of August, and at that time no perfect insects were as yet visible. The larvæ were of all sizes, some very small, and apparently but just hatched: here and there imbedded in the cottony substance were the eggs themselves. The former were rather active in their movements, of a green colour, with six rather short feet, the hinder pair not longer than the others; the antennæ also short, of six joints. The pupæ had rudiments of

* One of the genera belonging to the family of *Aphidæ*, or plant-lice, but distinguished from *Aphis* by its short filiform antennæ, tomentose body, and abdomen destitute of horns or tubercles towards the apex.

wings, but were similar to the larvæ in all other respects, except in being larger; they were exactly a line in length. On placing some of the lettuces under a bell glass, several of the perfect insects appeared on the 3rd of September; others following in succession for some time afterwards. These were of two colours, perhaps characteristic of the two sexes. Some had the head and thorax dusky brown; the abdomen pale dusky, tinged with greenish-yellow; the legs dusky, with the joints rather darker: others inclined generally to ochraceous-yellow, especially the abdomen, and the collar between the head and thorax.*

Amongst the larvæ at the roots of one lettuce I observed a single specimen of the larva of some other totally different insect, which appeared to be feeding upon them. This latter was vermiform, and much attenuated towards the anterior extremity, which was very protractile; it was of a pale green

* If the above be an undescribed species of *Eriosoma*, which is extremely probable, from the little attention which has been paid to the insects of this family,—it might be named *E. lactucæ*, and thus characterized:

E. capite et thorace fuscis: abdomine oblongo fuscescenti-ochraceo, vel viridi-ochraceo; pedibus fuscis, articulis saturatioribus.

Long. 1. lin.—*Hab.* ad radices lactucæ sativæ.

Possibly it may be the *Aphis radicum*, briefly alluded to by Kirby and Spence, (vol. ii. p. 89,) as deriving its nutriment from the roots of grass and other plants. There are, however, without doubt, several species of these root *Aphides*. I have occasionally observed another, besides the one described above, at the roots of the *Lysimachia nummularia*, when growing in a pot in my garden, and rendered unhealthy by being kept too dry. This was like-

colour, and about two lines in length. There were also some small brown coccoons among the roots, here and there, likewise about two lines in length, which I kept, in the hope of their turning to the perfect state, but without success. Probably these were the larva and pupa respectively of some dipterous insect, which keeps the root *aphis* in check. When once, however, the nuisance occasioned by this last parasite shews itself in a garden, the only effectual way of getting entirely rid of it is immediately to pull up all the diseased plants and burn them.

LICE.*

ALMOST every animal, as well as bird (it is well known), has its peculiar louse. It is, however, singular, and contrary to what one might have expected, that the size of the parasite is by no means, in all cases, proportioned to the size of the animal it infests.—The louse of the swine† is as large as the

wise a species of *Eriosoma*, but differed from the *E. lactucæ* in having the abdomen shorter and broader, (or more approaching to round than oblong,) and in being more sluggish in habit, hardly attempting to move when taken from the plant: it also kept more on the surface of the ground, at the bottom of the leaves and stems, than under ground; though many might be noticed at the roots themselves.

Reaumur has given a list of plants, at the roots of which he had found *Aphides*, but the lettuce is not included. *Hist. des Ins.* (12mo. ed. Amst. 1738,) tom. iii. 2nd part, p. 80.

* *Anoplura* of Leach.

† *Hæmatopinus suis*, Denny; *Anopl. Brit.* pl. xxv. f. 2.

louse of the ox : * the louse of the eagle † is not
larger than the louse of the rook ; ‡ while the louse
of the great snowy owl § scarcely exceeds that of
the tiny gold-crested wren.|| Birds also of the same
size harbour species that are very dissimilar in this
respect. The louse of the sparrow-hawk ¶ does not
measure more than three-quarters of a line in length;
while that of the hobby ** (of which I have two
specimens taken by myself from that bird) is among
the most gigantic of its race, and exceeds a quarter
of an inch.

Some species of birds are much more troubled
with these parasites than others. The rook and the
common buzzard appear to be particularly infested
by them. The latter I have often found swarming
with them, more especially the feathers about the
head and neck, to an extraordinary degree. Other
birds, again, appear to be nearly free from lice altogether. The common heron is one, on which I
never could detect a single specimen, though repeatedly searched for this purpose.†† It once occurred
to me, that the dust, which naturally abounds so in
the plumage of this bird, might have the effect of

* *H. eurysternus,* Id. pl. xxv. f. 5.
† *Docophorus aquilinus,* Id. pl. ii. f. 7.
‡ *D. atratus,* Id. pl. iv. f. 8.
§ *D. ceblebrachys,* Id. pl. i. f. 3.
|| *D. reguli,* Id. pl. vi. f. 4.
¶ *D. nisi,* Id. pl. iii. f. 11.
** *Læmobothrium laticolle,* Id. pl. xxiii. f. 4.
†† Mr. Denny, however, has described two species found on the
heron, *Lipeurus leucopygus* (pl. xiv. f. 4.) and *Colpocephalum im-
portunum* (pl. xviii. f, 1.), though the former would seem rare.

keeping them off; as it is generally supposed that the object for which common poultry and other birds dust themselves is that of ridding themselves from these vermin, or preventing their accumulating.

MITES.*

Tick (Ixodes).—WHILST examining a blue titmouse (*Parus cœruleus*), I was surprised to find under its feathers a specimen of a tick, nearly equal in size to the dog-tick,† and to all appearance identically of the same species. One would hardly expect to meet with so large a parasite on so small a bird.‡

On the same bird were large numbers of a very minute sort of mite, so small indeed as to be scarcely visible to the naked eye. I examined some of them with the microscope, and found to my astonishment that they consisted of several distinct species, referable in some cases to distinct genera. This led me to reflect on the numberless tribes of minute beings

---

* *Acaridæ;* thought by some to constitute a peculiar class among the *Annulosa*, but more probably only a subdivision of the class *Arachnida*.

† *Ixodes ricinus*, Leach.

‡ Dr. Leach, in a paper published in the eleventh volume of the Linnean Transactions, has described the British species of *Ixodes*, one of which (*I. pari*) is mentioned as found on the great titmouse (*Parus major*). This species I have myself taken from that bird; but it is quite different from the one noticed above, and much smaller.

that probably yet remain to be detected by the
naturalist. Perhaps there is no species of bird, certain individuals of which may not harbour in their
plumage hosts of these little animated specks, consisting of various kinds destined to feed on the
superabundant juices of the body. If so, how many
hundreds, one might almost say thousands, of species
of such *Acari* must there be of which we know
nothing! And though some may think it trifling
to bestow much time or trouble in the investigation
of these atoms,—yet it should be remembered that
our acquaintance with the entire scheme of nature
can only be advanced so far as we search into all the
varying forms of structure that occur, be the individuals in which they appear large or small.

*Sarcoptes passerinus.**—This is one of the more
common forms of *Acari*, parasitical on birds, and
remarkably distinguished by the disproportionate
size of the third pair of legs. It is well described
by Linnæus, who, in reference to this circumstance,
says,† *pedes tertii paris magni; femora enim hujus
ambo simul mole ipsum corpus fere adæquant, reliquis
minimis: his monstrosis pedibus non incedit, sed modo
sese adfigit pennis vivacissimæ et agilissimæ avis, ne
excutiatur.*—In confirmation of this last remark I
observe, that, when placed upon a piece of white
paper, this creature in walking draws these enormous
legs after it as a dead weight, without attempting to

* *Acarus passerinus* of Linnæus. It is well figured by De
Geer, tom. vii. pl. vi. f. 12.
† *Faun. Suec.* p. 480.

o

assist its march by them, which is entirely effected
by the other three pair. I suspect, however, that
there are several distinct species of these *Acari,*
tertiis femoribus incrassatis, which have been hitherto
confounded:* some that I have taken from the green
woodpecker are of a much more oblong form than the
more common sort figured by De Geer, with the
third pair of legs of more moderate size, and not so
enormously thick.

Acarus domesticus.†—The mites which infest
museums, and make such havoc in collections of
dried specimens, appear to me to be of two or more
kinds. One, however, seems identical with the
cheese-mite, as described and figured by De Geer;
at least, I can discover no difference between them
on the closest comparison. These little creatures
are of active habits, and run with considerable speed.
They appear to shift their skin very often, as appears
by the numberless empty sloughs that may be found
in their haunts. When first hatched, and for some
time after, like many others of this tribe, they have
but six legs, the two others not sprouting till they
are near full-grown. The deficient pair is always
the third.

Cheyletus eruditus.‡—This little mite, which seems
scarcely to have been noticed by English authors, is
remarkable for its very large and conspicuous sickle-

* They will probably be found eventually to constitute a
peculiar genus by themselves, subordinate to *Sarcoptes* of Latreille.

† *De Geer,* tom. vii. p. 88. pl. v. ff. 1-4.

‡ *Lam. Hist. Nat. des An. sans vert.* (2nd edit.) tom. v. p. 75.

shaped palpi, ending in a minute cheliform process. These palpi it moves about in a very free and active manner, constantly opening and shutting the cheliform fingers at the extremity. Its motions, in this respect, are very similar to those of the genera *Chelifer* and *Obisium* among the pseudo-scorpions. I have occasionally taken this species in collections of dried plants that had been kept in rather a damp place. Either it, or a closely allied species, is sometimes found parasitical on birds: such I have taken from the green woodpecker, differing only from the common kind in being almost quite white, the posterior extremity of the abdomen alone inclining to yellowish. None of these mites, however, have more than six legs, and possibly they may prove to be only the larvæ of some other genus.

*Philodromus limacum.**—This mite is parasitic on the larger kinds of slugs, especially the common black slug, on which it may frequently be observed in great plenty. The most striking feature in its history is the circumstance of its not confining its abode to the external surface of the slug, but often retiring within the body of that animal; effecting its entrance by means of the lateral foramen which leads to the cavity of the lungs. Indeed, I am inclined to think that this cavity is its principal residence, whence it only comes forth occasionally, to ramble upon the surface of the body. In one in-

* See *Loudon's Mag. of Nat. Hist.* vol. iv. p. 538, where I have given a detailed account of the characters and habits of this little mite, from which the above is taken.

stance, I confined in a close box a slug which, to all appearance, was free from parasites. On opening the box a day or two afterwards, I observed very many crawling about the slug externally, all of which would seem to have proceeded from the pulmonary cavity. On another occasion, I observed these insects running in and out of this cavity at pleasure; and some which I saw retire into it never reappeared, although I watched the slug narrowly for a considerable time. It is remarkable that the slug appears to suffer no particular inconvenience from these parasites, and even allows them to run in and out of the lateral orifice without betraying the slightest symptoms of irritation.

Another curious circumstance in the history of these little mites, is the extreme rapidity of their motions. One might have thought that so slimy a ground as the surface of a slug's back would have impeded their progress; but so far is this from being the case, that they are never to be seen at rest: indeed they run with greater celerity than any other parasites of this tribe I am acquainted with; these insects, in general, being rather slow in their movements than otherwise. From this circumstance, it becomes difficult to secure them for examination; and the more so, from their bodies being of a soft nature, and crushed by the slightest touch. Perhaps the most effectual way is to drop the slug into very weak spirits, sufficient to paralyse the limbs of the parasite, without immediately destroying life. If dropt into plain water, the mites rise to the top, and run about with as much ease and activity upon the surface of the fluid as upon the slug itself.

RED SPIDER.[*]

In the summer of 1844, the gardener at Bottisham Hall directed my attention to a peach-tree infested with the red spider. The tree had a very diseased appearance ; the leaves were red and spotted on their upper surface, and fell prematurely, stimulating the tree to put forth new leaves from the next year's buds, and so subjecting it to great exhaustion. On examining the leaves through a microscope, they were found to be infested with a very minute mite, scarcely bigger than the point of a pin, of an oblong-oval form, with eight legs of not very disproportionate lengths, but the anterior pair a little the longest ; the body of a yellowish orange-colour, some specimens redder than others, and some with an oblong dark spot on each side of the abdomen ; the legs pale whitish, and transparent ; two eyes, one on each side of the anterior part of the body, forming a bright red spot. These mites were in all stages of growth, and had evidently been bred upon the leaves. The young were more pale-coloured than the adult, which were often deep red ; the former were also much more bristly than the latter, the hairs seeming to wear off with age. Here and there might be discerned (more especially on the

[*] *Acarus telarius,* Linn.; *Tetranychus telarius,* Duges, *Ann. des Sci. Nat.* (2nd ser.) tom. i. p. 25.

young leaves that were prematurely forced, as above mentioned,) clusters of minute globules that appeared to be eggs.   There were also, on the tree, great numbers of *aphides*, which appeared to have had some share in bringing it into its diseased state; and also a few of the larvæ of what was supposed to be a *Hemerobius*, preying upon both mites and aphides, empty skins of which were scattered everywhere upon the leaves.

This mite is no doubt the *Acarus telarius* of Linnæus, so called from a thin filmy web which it spins, and with which it mats the under side of the leaves of the trees it frequents.   Such a web I discovered upon the leaves in question; but the threads of which it is composed are so extremely fine, that, when single, they can be with difficulty perceived, except under a high power of the microscope.   The web alone must injure the health of the leaves, independent of the punctures made by the mites themselves, which last were distinctly seen in some instances inclining forward the anterior extremity of their bodies, and plunging their rostrum into the leaf to feed.   The mites run upon the threads of their web like spiders, but not very swiftly; off the threads their motions seem impeded, and upon a smooth surface like glass it is with some difficulty they move at all.

When once a tree has become much infested with this parasite, its cure is next to hopeless.   In the present instance, the tree had shewn the disease year after year; and there was nothing to be done, as I was told, but to cut it down, which has since been effected.   Some have recommended frequent sprinklings with cold water,* at least for plants in hot-houses, which are often attacked by these insects; but this remedy is not a certain one, nor to be relied on as permanent in its effects. In the case of fruit-trees labouring under this disease, the leaves are constantly falling, and carrying with them to the ground numbers of the mites, which either ascend the tree again, or, if it be late in the season, harbour under stones, &c. near its foot, till the ensuing spring.

CHELIFER GEOFFROYI.†

I ONCE took four specimens of this insect from one little fly, which was so fettered in its movements by the grasping hold of so many aggressors at once, as hardly to be able to crawl. The natural habitat of these pseudo-scorpions appears to be under the bark of trees, whence they probably spring out on their unwary prey basking near. They are often found on *Tipulæ*.

ENTOMOSTRACA.

Cyclopsina castor.‡—THIS little crustacean is very amusing to watch in its habits. It occurs in pools of water in this neighbourhood not unfrequently.

---

* See *Kollar's Treatise on Insects injurious to Gardeners, Foresters, and Farmers*, by Loudon, (p. 182,) where there is some notice of this insect.

† *Leach, Zool. Misc.* vol. iii. tab. 142, fig. i.

‡ *M. Edwards, Hist. Nat. des Crustacés*, tom. iii. p. 427.

It often remains at rest in the same spot for several seconds together, with its antennæ extended and motionless, but its feet in a state of intense vibration, and then after a while suddenly darts off to another situation with wonderful rapidity. It would seem also to be tenacious of life; as I observed that, amid a large variety of different *Entomostraca* that were kept in a vessel of water, which was not changed for a considerable time, it survived all the others by several days.

Argulus foliaceus.*—In 1835, I found this singular species, which has been very little noticed by English naturalists, in fish-ponds at Ely in great abundance. They are there parasitical on pike, adhering to different parts of the body of that fish. The fishermen called it the *pike-louse*.

ELECTRIC CENTIPEDE. †

Dec. 10th, 1843.—A MAN brought me to-day what he called two glow-worms, which he had seen shining on a bank the preceding evening. They proved to be only the electric centipede, which is frequent in the autumnal months in this neighbourhood, and may often be seen shining by road-sides, more especially on mild damp evenings. They are constantly mistaken for glow-worms by the common people.

---

* *Hist. Nat. des Crustacés,* tom. iii. p. 444.

† The *Scolopendra electrica* of authors; but probably under that name two or more species have been confounded, as I suspect there are at least two in this country which are luminous. It belongs to the genus *Geophilus* of Leach.

These two individuals I kept alive in a small box with damp earth at the bottom, for several months; but I never observed them to shine once after they were taken.*

The luminosity of this insect is remarkable for diffusing itself over the ground, or the fingers of those who attempt to handle the individual from which it emanates, in like manner as when pure phosphorus is handled. I am not aware that this is ever the case with the real glow-worm.

* I was not then aware of the observations of Mr. Gosse, who states that he found the light might be immediately reproduced by breathing upon the centipede: the experiment fails (it is said) if repeated too often at once, though after the lapse of a day and night it will produce the same results as at first. See *The Zoologist*, p. 160.

OBSERVATIONS ON WORMS.*

THE different kinds of worms, as they may be called in a general way, inhabiting the land or fresh-water, appear to be numerous: yet they have received but little attention from naturalists, at least from the naturalists of this country. They whose inclinations have led them to the study of these tribes, have for the most part restricted their attention to the marine species, which doubtless are more attractive from their larger size, and greater variety.— Nevertheless, those abovementioned are not without interest, and would repay a careful investigation of their characters and history. Those found free, that is, not inhabiting the bodies of other animals, belong principally to the old genera of *Lumbricus, Nais, Gordius, Hirudo,* and *Planaria,* which have been since divided into several others. Some of those common in ponds, though not strictly parasitical, are often found infesting the shells of certain fresh-water molluscks, more particularly the *Physa fontinalis, Paludina vivipara,* and *Planorbis marginatus,* which appear particularly to suffer from their attacks, wriggling much about in the attempt to rid themselves of their enemies, which creep round them, and adhere very closely to their soft parts. It seems

* Comprising the two groups of *Annelida* and *Entozoa* of modern naturalists.

not improbable that some of the minuter worms of this kind, including one which is frequently to be met with in wet moss, and soft decomposing wood that has lain long in wet places, may be the first forms of other species which have been hitherto considered distinct. But as yet we are very much in the dark respecting the modes of propagation and development of these obscure animals, and scarcely know what changes they may pass through before arriving at their full growth:—this is especially the case with the *Entozoa*.*

SMALL RED DITCH-WORM.

THERE is a small red worm, about an inch and a half long, not uncommon in this neighbourhood, in ditches, and in the rills which are cut to drain pastures, either burrowing in the mud, or concealing itself amongst aquatic plants, particularly the *Chara vulgaris*, to which it shews much attachment. It has not often been noticed by English naturalists, but it appears to be the species described by Müller under the name of *Lumbricus variegatus*.† This worm has often astonished me by its powers of reproduc-

* Since the above was written, I have seen the important memoir by Steenstrup, *On the Alternation of Generations*, an English Translation of which forms one of the last publications of the Ray Society. The observations of this author, if correct, throw a new aspect over the whole subject of the origin and first life of the *Entozoa*, as well of some other of the lower animals.

† *Vermium Historia*, vol. i. pt. ii. p. 26. See also *The Zoological Journal*, vol. iii. p. 326, where it is well described by Dr. Johnston, the only English author who, so far as I am aware, has noticed it.

tion and multiplication, when divided into the smallest pieces. Müller, indeed, has noticed this property which it possesses; and states it moreover to
be the species which was the subject of Bonnet's
experiments, which are well known.* But what has
most struck me, is the readiness with which a severed
portion will acquire a fresh growth of parts, both
anteriorly and posteriorly, not merely when placed
in the same ditch-water in which it ordinarily lives,
and which may be conceived replete with animal and
vegetable matter, but when kept in pure pump-
water, renewed every day. Such water, when quite
fresh, rarely, if ever, is found to contain the smallest
animalculæ, still less anything of a vegetable nature;
so that one wonders whence the worm, under such
circumstances, can derive nutriment sufficient, not
merely for maintaining life, but for enabling it to
reproduce lost parts. How great must be its powers
of assimilation to elaborate a new head and tail from
the purest water in the course of a few days! The
smallest portion, cut out of the middle of a perfect
worm, is not long in doubling its own length in each
direction. With the view of ascertaining what the
limits of its powers were in this respect, I, on one
occasion, put such a portion into distilled water, but
found that when thus treated it speedily died, without acquiring any fresh growth.

This little worm is at all times of so delicate a
structure, that, after being removed from its native

* Expériences sur la reproduction d'une espèce de Vers d'eau
douce. Œuvres, tom. vi. p. 20. (8vo edit. Neuchatel, 1789.)

ditches, the slightest touch is sufficient to cause the body to separate into pieces.

EARTH-WORM.*

In the month of November 1842, my nephew found a hole, in a waste piece of garden-ground, similar to, but rather larger than, a rat's hole ; it penetrated three feet into the soil in a slanting direction, and at the extremity, about nine inches from the surface, a quantity of large earth-worms were discovered, about as many as could be held in both hands. Some of the worms were alive; but others were dead, and appeared as if they had been bitten in two. There were also, with the worms, a small collection of dead leaves, forming a kind of nest.

Query, whether these worms had been stored up by moles to serve for their winter's provision ;† or whether, from the circumstance of the long hole leading to the deposit from the surface of the ground, it be not more probable that they were amassed by some other animal ?

White has remarked that " earth-worms make their casts most in mild weather about March and April." These casts, however, are found at all seasons : I observe that even throughout the winter, if there be no frost, they reappear as often as the lawn is swept.‡

* *Lumbricus terrestris*, Linn.

† Mr. Jesse has mentioned, on the authority of a mole-catcher, that moles are in the habit of laying up worms in this manner before winter sets in. *Gleanings* (2nd series), p. 26.

‡ I may take this opportunity of drawing the attention of

CLAY-WORM.

March 20, 1826.—A MAN brought me to-day what he called a *clay-worm*, which he threw up with his spade while digging. It was rather more than ten inches in length; filiform, about the thickness of a horse-hair; and white, inclining to yellowish; slightly attenuated towards the tail. The head was obtusely conical. Scarcely any trace of a mouth could be discerned, except under a very high magnifier. Whether

English readers to a fact said to have been observed by a foreign naturalist, which I have never noticed myself, but which it would be very interesting to have confirmed by the observations of others. It respects the phosphorescence of a certain species of earth-worm during the act of coupling. The following passage is translated from a French Journal, in which the fact is recorded.

"A great number of small phosphorescent animals were noticed, two or three years back, in a garden at Toulouse, during a very warm summer's evening. MM. Saget and Moquin-Tandon, having carefully examined them, recognized them as positively belonging to the genus *Lumbricus*. They were from forty to fifty millimetres* in length, or thereabouts. The light they emitted appeared bluish, and much resembled that of iron at a white heat. When one of these worms was crushed with the foot, the light spread itself on the ground; it could be made to leave also a long luminous track, as if one had rubbed the ground with a piece of phosphorus. Each of these *Lumbrici* shewed a *clitellum* well developed, which proves that the individuals observed were adult, and in the act of coupling. M. Moquin-Tandon observed that the luminosity resided in the substance of the *clitellum*, and that this property ceased to exist immediately after coupling." *L'Institut,* tom. viii. (1840) p. 381.

---

* Equalling rather more than an inch and a half to two inches, English measure.

this be the *Gordius argillaceus* of Linnæus is perhaps doubtful; but it is quite distinct from a worm
somewhat similar, but much thicker, as well as of a
darker colour, occasionally found in water, which last
I suppose to be the *Gordius aquaticus.**

THREAD-WORMS.

A REMARKABLE phenomenon occurred at Fairford in Gloucestershire in June of the last year
(1845). On Sunday the 15th of that month, countless numbers of thread-worms, apparently allied to
the genus *Filaria* † of naturalists, were found in a

* It is also quite distinct from the *Thread-worms* next noticed.
It is probably a species of *Filaria*, but I have not as yet satisfied
myself on this head.

† The worms alluded to in the above notice do not belong to
the genus *Filaria*, but to that of *Mermis* of Dujardin. The
reader is referred to a memoir by that naturalist, published in the
Annales des Sciences Naturelles for 1842, (tom. xviii. p. 129,) in
which he has given the distinctive characters, and what he considers to be the true history (so far as he has yet made it out),
of this new genus of worms, closely resembling both *Filaria* and
Gordius, and hitherto confounded with this last by some observers. They are worms, which, it would seem by his account,
have been not unfrequently noticed, sometimes in great abundance,
on the moist ground after rain, or the morning after a strong
dew, sometimes also on newly-dug borders, or on the box edging
of borders, twisting themselves round the plants like whitish
threads spotted more or less with black internally. He believes
them to be the same as what Goeze speaks of as having found in
the month of June spread over the newly-dug borders of a garden by hundreds after a heavy storm of rain, so long back as the
year 1781, but which he confounded with the *Gordius aquaticus.*—

garden in that town, after an exceedingly hard storm,
and under circumstances that at first sight seemed to

He believes also that it was these worms which Audouin found in
considerable quantity in the larvæ of the cockchaffer, of which
Dujardin conceives them to be the parasites, during the first stage
of their existence. And this serves to explain why the worms
have been only met with in certain seasons, as appears to be the
case. He supposes that, the larvæ of the cockchaffer being
many years in coming to their full growth, their intestinal worms
would, like them, shew themselves more abundantly some years
than others. — We may explain also (he adds) why it is generally
on the surface of the earth in gardens, where these larvæ live at
a certain depth, that the Mermis shews itself in the spring, when
the earth has been disturbed, or softened by rain : their appearance is furthermore occasioned by the moisture stimulating the
larvæ already sick to expel their parasites by contraction, and
thus allowing the Mermis to get free, in order to deposit its eggs
upon the ground, without being dried up.

Dujardin is of opinion that many of the long filiform worms,
reported to have been found in the bodies of different caterpillars
and insects, have been really species of Mermis, the colour of
which, becoming more and more dark according to the degree of
development of the ova, has led to their being confounded with
the true *Gordius*, as well as the circumstance of their being able
to live a considerable time in pure water.

He has himself kept the Mermis in water for more than eight
days ; but he has always observed that the deepest-coloured individuals tried to escape from the water, with a view of depositing
their eggs in dry places, where they soon died. When the worms
are not pressed by this necessity, he states that they can subsist
for a much longer time in pure water; and this is precisely the
case with those that have been drawn from the bodies of larvæ or
living insects.

The above is an abridged extract from Dujardin's account of the
habits of these worms. He has followed it up with a very
detailed description of their anatomical structure and the cha-

indicate that they must have been precipitated with the rain. I am indebted to Sir Thomas Tancred, of

racters by which they are distinguished from those genera of worms most nearly allied to them. He has dwelt especially on the tegumentary system, and the reproductive organs, which last have much of peculiarity in respect of the mode in which the eggs are gradually developed. The digestive apparatus is very simple, at least in the adult state; for the embryo is provided with a distinct intestine. And comparing this embryo, (so similar to the small eel-like worms found in damp earth, and amongst moss, as well as in the visceral cavities of earth-worms, slugs, insects, and their larvæ,) with the adult Mermis, in which, by reason of the excessive growth of the tegumentary system and the reproductive organs, the digestive apparatus is become incomplete and in part rudimentary, Dujardin is led to conclude that the Mermis, as already observed, is the last stage of development of an entozoon, different from all the known *Nematoideæ*, undergoing great changes with age, and only arriving at this last stage, in the bodies of larvæ or insects whose life is sufficiently long. It comes afterwards into the air merely to lay its eggs, and then dies. It is not, therefore, a terrestrial or an aquatic worm, but an intestinal one; for which, on account of its peculiarities, Dujardin conceives it necessary to create a new order, intermediate between the *Nematoideæ* and the *Acanthocephali*.

The characters of the genus *Mermis* are thus laid down :— *Vermis, corpore longissimo filiformi, elastico, anticè parumper attenuato ; capite subinflato, ore terminali minimo rotundo ; intestino simplice, posticè obsoleto, ano nullo ; vulvâ anticâ, transversâ.*

Ova juxta placentas lineares, intra tubum muscularem concepta, denique in capsulis monospermis, bipolaribus, bipedicellatis, deciduis inclusa.

The species described by Dujardin, and which there is great reason to believe is identical with the worms noticed in the text, and found at Fairford, he calls *Mermis nigrescens*. It is thus characterized : — *Mermis caudâ obtusâ, capite subangulato ex papillis 5-6-obsoletis ; ovis nigris.*

Stratton, in that county, for the first information I
received of this phenomenon, which seemed so singular an occurrence, that, on a late visit into Gloucestershire, I went to Fairford myself, and called upon
Mr. Samuel Vines, of that place, the individual by
whom it had been noticed, in order to examine the
spot, and obtain further particulars.   The result of
my inquiries quite satisfied me as to the accuracy of
the facts reported to me, which I proceed to detail.
Mr. Vines informed me that there had been an extremely heavy rain at Fairford, accompanied by
thunder, about the middle of the day; and that a
little earlier there had been even a water-spout at
Castle Eaton, a small village situate about three
miles south-west of Fairford, from whence the storm
came ; being, as it would seem, the residue of the
water-spout in some measure spent before its arrival
at the place last-mentioned.   On going into his garden immediately after the rain had ceased, between
the hours of three and four in the afternoon, he first
observed some of these worms on the top of the box
edging by the side of the walk; he afterwards observed others, both on the soil of the borders, and
on the walk itself; and, on carrying the search further, soon discovered that they were scattered over
the whole of the garden, three or four on the average being to be met with in every square foot.   He
next noticed that they were not confined to the
ground, but that they were equally abundant on the
leaves of the potatoes, and other crops of vegetables
in the garden; and even upon the tops of the fruit-
trees, as high as he examined, certainly seven or

eight feet from the soil. In short, wherever he sought
them, they were to be found; and in his opinion there
must have been thousands altogether. When first
noticed, the worms were alive and writhing : those
which lay upon the ground did not attempt to go
into it, but remained on the surface till dried up by
the sun; and this was speedily the case with many
that were observed upon the flag-stones immediately
before the house; but others, that were on the box
edging, and found shelter amongst the leaves of the
plants, retained their vitality for a longer period, and
reappeared the Tuesday following, when there was
another fall of rain at Fairford, though not in any
great quantity.* On placing some of the worms in
water, soon after they were obtained, they did not
shew much activity in their motions, but twisted
themselves into knots, and remained at the bottom
of the vessel in a kind of half-torpid state, for the
period of about fourteen days; at the end of which
they died, notwithstanding the water had been occasionally changed during the time. On inquiry I
found that a few of these worms had been seen also
in the neighbouring gardens, but only on the ground;
it did not appear, however, that they had been looked
for elsewhere, nor that they had much attracted the
attention of other observers.

Mr. Vines was kind enough to give me a number
of these worms, which he had preserved in spirits of
wine. They were very slender, of the thickness of

* It appears, however, that there was an unusually heavy fall of
rain at Oxford on this day.

a fine thread, and measuring from three and a half to six inches in length; somewhat attenuated anteriorly, with a slight contraction behind the head, which was oblong and a little truncated. The tail was obtuse, and rather narrower than the body. They were of a whitish, or yellowish-white colour; transparent at first, according to his account, but rendered nearly opake by the spirit when I saw them. Just at the contraction behind the head was a reddish stain. There were several dark spots and streaks about the middle of the body, and a few nearer the extremities, which under a high magnifier were found to be the ova, apparently quite ripe, and ready for exclusion. These ova were of a deep brown colour, and were very abundant and distinct in some individuals. The body was cylindrical and quite smooth in the moist specimens; but in one, which had been suffered to dry from the living state, the sides of the body, especially about the middle, presented a number of transverse plaits or folds, so as to shew a crenated appearance: these, however, disappeared after some hours' immersion in water, when the body resumed its natural appearance.

These worms undoubtedly belong to the genus *Mermis* of Dujardin, to whose memoir, giving a detailed account of their structure and organization, I have referred in a note. They accord exactly with his description, and are in all probability identically the same species, which had been noticed by him and others under circumstances very similar to those above recorded. The only

fact of importance, in addition to what he has
mentioned, is that of their having been found in
this instance upon the trees in the garden, as well
as upon the ground and low vegetables. Dujardin
mentions their occasional appearance in large numbers in gardens after rain, both on the soil and
on the box edging so common by the side of borders, and the theory by which he accounts for it
is stated below. But to explain the circumstance
of their being met with at Fairford on the tops
of the trees, consistently with this theory, we
must either suppose that the worms, after their
exit from the soil, and when searching for a spot
to deposit their eggs, actually ascended to the height
at which they were found, working their way up
the stems of the trees by a continued writhing, and,
as it would seem, in a marvellously short space of
time ; or that some of them had not quitted the
bodies of the grubs in which they were originally
imprisoned till after these last were transformed
into beetles, and had conveyed themselves to the
trees.* At the same time, it is possible that some
of the worms, which had climbed to so considerable
a height, might have first left the ground on a
former occasion, only not have been noticed till
after the hard rain on the 15th which caused them
to revive ; in like manner as those which were
found on the box edging reappeared two days after
on the return of wet, having in the mean time

* I did not hear, however, that any cockchaffers had been
observed in the garden previously.

concealed themselves amongst the leaves. Also, when attention had been attracted towards them by the occurrence of others in such numbers on the ground, they might readily be perceived, when but for this circumstance they would have passed unobserved. The fact of their having been obtained at such elevations is quite certain. Mr. Vines was so good as to take me into his garden, and to point out the exact spots in which he found them; and everything he stated was confirmed by the testimony of a second person, who was employed as a labourer in the garden and at work when I called. Possibly such a phenomenon may be more frequent than is supposed; though I am not aware of more than a slight notice of it, in one previous instance, in this country.* If so, the recording of it in this instance may have the effect of leading naturalists to look out for, and to discover, other instances of its occurrence in their own neighbourhood. They will also be directed by the foregoing details to the particular points requiring further examination; and be en-

---

* " On Tuesday, June 19, after a very heavy storm, an extraordinary phenomenon was observed at the Buckhold Wood, Lydart, Troy Park, and other places in the neighbourhood of Monmouth. The ground and trees were covered with myriads of live snake-like insects, quantities of which were collected by different persons. They are six or seven inches in length, about the thickness of a horse-hair, white, and quite transparent." *Mag. of Nat. Hist.* vol. ix. p. 241, quoted from the *Literary Gazette* of June 23, 1832. The worms in this instance were erroneously referred to the *Gordius aquaticus.*

abled perhaps to remove so much as there is of
mystery still enveloping the whole subject.

WORM IN A GNAT.

EVEN the smaller insects do not appear exempt
from worms, which sometimes are of such a size,
compared with that of the species they infest, as
to occupy almost the entire cavity of the abdomen.
I once caught a gnat* by the side of a stream,
which, shortly after it had been secured, gave
birth, from between the segments of the abdomen,
to a worm more than an inch in length. This
worm was thicker than the thread-worms noticed
in the last article : when it was wholly out of its
prison, the abdomen of the gnat collapsed like an
empty glove.

VINEGAR EEL.†

THE little eels found occasionally in vinegar are
well known. Some persons suppose that they are
to be found at all times in that liquid, when
sought for with the microscope. This, however,
is not the case. It is only vinegar, which has
been kept some time, and which is become bad,
that gives birth to them. What is more, they will
not live long in good vinegar. This I ascertained

* *Chironomus plumosus* of Meigen, or a nearly allied species.
† *Vibrio anguillula,* Müll.? *V. aceti* of some authors.

on one occasion, when, on adding fresh vinegar to
some that had been given me by a friend full of
these eels, their motions from that day became
more and more languid, and in a fortnight's time
they were all dead.   Previously to this, I had
kept them several weeks in the same vinegar in
which they had been brought to me.   When these
little eels are immersed in sweet oil, or proof
spirit, their motions are much impeded, but vitality is not immediately destroyed.

LIVER FLUKE.*

THIS parasitic worm is well known to be the cause
of *rot* in sheep, infesting the liver and biliary ducts
of that animal.   A few individuals are to be met
with in most sheep; but their presence seldom leads
to any fatal results, except when they occur in great
quantities.   Such was the case in 1828; in the autumn of which year, and during the ensuing winter,
the above-named disease prevailed generally in low
grounds in this neighbourhood, and in some of the
more marshy districts in the county raged to an
alarming extent.   The previous summer had been
very wet, heavy rains continuing to fall without
intermission throughout July and August.

According to the statement of the farmers in these
parts, sheep contract this disease by feeding upon
the long rank grass, which springs up abundantly

---

* *Fasciola hepatica,* Linn. *Distoma hepaticum* of modern
helminthologists.

in pastures that are low and swampy, or which are
subject to inundation. They say that it is chiefly
after a wet summer or autumn that the disease
shews itself; that on such occasions, particularly
in marshy districts, it generally prevails to a greater
or less extent; but some think that if the rains do
not abound until the winter's frost is over, the sheep
are not so liable to it. Feeding sheep for a very
short time in wet pastures is sufficient to bring
on the disease: a relation tells me, that, some
years back, his flock contracted the disease by only
one day's feeding upon a wet moor. This account,
on the whole, seems to favour the opinion that
these worms are bred in the water, and adhering
to aquatic grasses, either in the egg, or, as it may
be termed, the larva state, (in which last state they
may possibly differ greatly from the adult fluke,)
are swallowed by the sheep.* The greater part
of these worms may perhaps be destroyed by the
frost of winter, which would serve as a reason
why there are less apprehensions of this disease

---

* The following case, recorded by Dr. Watson, seems to confirm the idea of these worms being swallowed by the sheep in the
first instance, in some state or another:—"A healthy flock of sheep
were driven through a considerable tract of country, and one of
them on the way broke its leg, and had to be carried on horse-
back. For one night the flock, with the exception of the maimed
one, rested in a marshy meadow, and every individual was seized
with the rot but itself; it escaped the disease, and had no liver
fluke. May it not be assumed, that the flock swallowed the eggs
of the fluke with the fodder they cropt from the moist meadow?"
I have copied this from the volume of *Reports on the progress of
Zoology and Botany*, recently printed by the Ray Society, p. 291.

in the early part of the spring, however wet that season may be.

This disease does not appear to be confined to sheep. I understand that hares in particular are very subject to it, and that they not unfrequently die in great numbers the same seasons in which it prevails among sheep. This was the case during the winter of 1828-9, above referred to, when many hares were picked up dead in their forms, presenting on dissection all the appearances that sheep exhibit from the effect of this complaint.

On throwing some of these flukes into cold spring-water, immediately after they had been taken from a sheep's liver, they contracted their bodies, and shewed some slight symptoms of life, but very soon died. This, however, is no argument against their aquatic nature in an early stage of their existence, as great changes may possibly take place in their organization and habits during their progress to maturity, similar to what we are familiar with in the case of many other animals which undergo meta-morphosis.

There is a good description, illustrated by figures, of this entozoon, in Hoole's edition of Leeuwen-hoek's works, with notes by the translator. (Pt. ii. p. a—h. cum tab.)

GROUND FLUKE.*

THERE is a small animal, somewhat resembling a liver fluke in shape, but more like a leech in its motions, common in this neighbourhood in damp woods and plantations, secreting itself under stones and rotten sticks. Of this animal, I can find no notice whatever by British naturalists. It is, however, most accurately described by Müller, under the name of *Fasciola terrestris,* which I have ventured to translate *ground fluke,* as an English name for this apparently neglected species. Its general appearance is more like that of a *Planaria* than any thing else I am acquainted with; and to that genus, as adopted by modern authors, if it be not a distinct one of itself, it is probably to be referred. It so closely resembles, indeed, some of the *planariæ* common in stagnant waters, that I have sometimes thought (especially from the circumstance of its often occurring on the banks of ponds, under stones resting upon the soft mud,) that it was only one of that tribe, which had been left on land by the partial drying up of the waters. But I have never noticed it in such situations, except where the ponds have occurred in woods; and in certain extensive plantations at Bottisham Hall it is so generally distributed, occurring where no water ever stagnates at all, (though the soil is of course damp, as it always is amongst trees,) that I am now quite convinced it is strictly a terres-

---

* *Fasciola terrestris,* Müller, *Hist. Verm.* vol. i. pt. ii. p. 68.

trial animal; though it cannot exist without much moisture.

The ground fluke is from half to three quarters of an inch in length. Its form is an elongated oval, when the body is stretched out for motion: but the latter is capable of great contraction as well as extension; and, when at rest, is gathered up into a small compass, as in the leech. Above, it is slightly convex; beneath, flat. The skin is viscid and opake, and of a dusky or greyish colour, darker in some individuals than others. When the body is contracted, it appears slightly annulose, or the surface compassed by fine ring-like striæ. The anterior part can only be distinguished from the posterior when the animal is in motion. There are two small black eye-like points at the anterior extremity, but they are not discernible, except when the animal is on the stretch, so as to attenuate the skin; and not then without the help of a strong lens. As the animal advances, it leaves a slimy track behind it, like a slug; and (as Müller observes) it might easily be mistaken for a young slug at first sight.

Müller mentions having met with this creature in September and November. I have observed it principally in the spring, but believe that it is to be found at most seasons of the year, when there is sufficient moisture in the soil and surrounding atmosphere to bring it into activity. I have not been able to satisfy myself entirely in respect of its food; but I am strongly inclined to believe that it is carnivorous, feeding principally upon the smaller land mollusks, especially those species which inhabit the same damp situations as itself. I have frequently noticed empty shells of the *Vitrina pellucida* in particular, also of *Helix nitida* and *nitidula,* as well as some others, near its abode; which is the chief ground of my thinking so.

I have occasionally kept these little animals for a short time in wet moss, placed in a box with damp earth at the bottom; but, if a certain quantity of moisture be not constantly kept up, they speedily die.

OBSERVATIONS ON MOLLUSKS.

FEW things are more remarkable in Natural History than the sudden appearance of species, in great plenty, in places in which they had been previously unknown. This has often been observed among insects, but such an occurrence is not confined to that class of animals. It happens not unfrequently with animals of other classes. I have twice especially had my attention called to this circumstance in the case of the fresh-water mollusks.—The first instance occurred in 1822. During the spring and summer of that year, some small pits in this parish, the bottom of which consists of a gravelly clay, and which are generally full of water, but sometimes dry, or at least many of them, during the warmer months, swarmed with the *Limneus glutinosus* of Draparnaud * to such an extent that the shells might be scooped out by handfuls: in some places, if a bucket had been lowered into the water, it might have been drawn up half full with them. Many other species

* *Amphipeplea glutinosa,* Nills.—Gray's edit. of *Turton's Man. of Brit. Land and Fresh-water Shells,* p. 243, tab. ix. fig. 103.

of *mollusca* were in company with the above *limneus;* but this species was the most abundant, and, from the circumstance of its being usually accounted rare, the most interesting of all. Many of the specimens were very large, much exceeding in size any I have seen in collections. These shells, however, did not prevail in any great numbers after that year. A few continued to shew themselves for three or four seasons, but they gradually disappeared; and now many years have elapsed since I noticed even a single individual.*

A somewhat similar phenomenon, in the case of another species, occurred in February 1825. The early part of that month had been very wet, causing the water to stagnate in large puddles in several parts of the park at Bottisham Hall, but which parts are not usually flooded, though sometimes a little swampy. Happening shortly afterwards to cross the park with a shell-net in hand, I immersed it into one of these puddles casually as I passed, when, to my surprise, I drew it out full of the *Aplexus hypnorum,*† a species which I had not at that time taken before in Cambridgeshire, though I have since met with it in one or two places. In this puddle the shells were collected in immense quantities, whilst none of the other puddles contained one. The shells were of various sizes, though none were full-grown. It were almost vain to speculate as to how they came there.

* I observe that Gray, in his *Manual,* speaks of this species as *periodically* abundant in certain places, which seems to be confirmed by what is above stated.

† *Gray's Manual,* p. 255, tom. ix. f. 113.

Even supposing that the spawn had been dormant in the soil, or conveyed there in any way the imagination can suggest, still how could the shells have acquired so rapid a growth in the short time the water had been standing in that spot?—The puddle was scarcely more than three feet by two across; it had not been in existence above a fortnight at longest; it was only a few inches deep; and half a dozen fine days would have been sufficient to lay it dry again. Such, in fact, proved to be the case before the month had expired, and the species has not been observed since in that locality.

———

Professor Henslow has communicated to me the following case of a mollusk being found in plenty in rather a singular situation, and in which also, as in the case last-mentioned it is not very easy at first sight to account for its appearance, though the conjecture he has offered is perhaps the most probable one.

"The Messrs. Ransome, at Ipswich, supply water to a trough, connected with the supply of a powerful steam-engine, which is obtained by a bore of considerable depth through the chalk. After sustaining a temperature of about 100°, this water is pumped into a large cistern upon the roof of their establishment, where it retains a temperature of about 70 degrees. This tank is regularly cleaned out once a year. My attention was directed to the fact of its being continually replenished with a very large number of the *Limneus pereger*. Individuals of small dimensions (when I inspected the tank) were

crowded over the sides, and many full-grown specimens were also crawling about. As the water from
the bore was conveyed pretty directly to the cistern,
it was supposed that these mollusks might have been
brought up with it; and that they had resisted the
effects of the somewhat high temperature to which it
had been subjected. But a little investigation suggests the probable method by which the spawn of
the mollusks is introduced. Some moss and confervæ
also grow in the cistern, and their sporules may very
possibly have been wafted there by the winds. But
the most probable method by which the spawn of
mollusks would arrive at this cistern, thus placed on
the top of the house, is (I conceive) by its becoming
entangled about the bodies of water-beetles, and I
observed three or four species of these tolerably
abundant in it."

HELIX LAPICIDA.*

Dec. 8th, 1822.—WE found a pollard elm this
morning pierced in all directions by the *Helix lapi-
cida*.—On tearing away the bark, and portions of
the wood, great numbers were observed of all sizes.
The tree was so weakened by their attacks, and so
much of its substance gone, that a slight wind would
have been sufficient to overturn it. These animals
appear to eat their way along in the same manner as
woodlice, and will soon devour all the internal wood
of a tree where they abound. Some of the indivi-

* *Gray's Manual*, p. 140, pl. v. f. 51.

duals observed in this instance were in a torpid state, and had stopped up the mouths of their shells with a bung of sawdust and small chips of wood cemented together.

HELIX CARTHUSIANA.*

IT is curious to observe any marked difference of habit between nearly allied species. Some years back, when collecting specimens of *Helix carthusiana* on the south coast between Dover and Folkestone, where this shell occurs in great plenty, I noticed that on taking them up, and holding them between the fingers for a few seconds, the animal, instead of retiring within its shell from alarm, as is generally the case with the *helices* when handled, invariably protruded itself to its full length, remaining extended so long as it was kept in that position. I never observed this in the case of the *Helix cantiana*, or of any other of those species most nearly allied to the *Helix carthusiana*.

CYCLOSTOMA ELEGANS.†

Jan. 8th, 1827.—WHILST searching to-day for fossils in Reche chalk-pits, we stumbled upon a few dead shells of the *Cyclostoma elegans*, which lay scattered over the rubbish that had weathered off from

---

* *Gray's Manual*, p. 146, pl. iii. f. 27.—*H. carthusianella*, Drap.
† *Id.* p. 275, pl. vii. f. 75.

the perpendicular sides of the pit. Upon looking further, a large number of specimens were found dispersed in all directions. The shells were of every gradation of growth from very young to full-sized. They were scarcely at all altered, the outer surface still retaining in part its usual colouring. They were also for the most part extremely perfect, even to the preservation of the opercle, which, in the case of many individuals, still closed the aperture of the shell.—It is a matter of curious inquiry how these shells came there, as the species is not known to occur anywhere in the neighbourhood. Though I have repeatedly searched for it in the living state, I never met with it in the entire county of Cambridge-shire, except in one locality near Linton,* and there only obtained a single specimen. The shells at Reche cannot have been brought from any distance, nor can they have remained long in their present unoccupied state. We can only suppose that they must have existed at no very remote period, if they do not still exist, somewhere in the immediate vicinity of the above pits, into which they must have been washed with the soil by the agency of some land-flood. Possibly they were confined to a small district, so that the entire race may have been drowned.

* Pennant, in his *British Zoology*, mentions its being found at Madingley Wood, but I never observed it there myself.

SMALL FRESH-WATER BIVALVES.*

SOME years ago, when studying these little animals,† I was greatly struck with the differences of habit observable in species, which externally so much resembled each other as hardly to offer, at least to an unpractised eye, any distinguishing characters.‡ These bivalves are common in ponds and ditches, as well as in streams; the smaller species are not unfrequently found in the gullies cut in the pastures for the purpose of draining the soil. They all live readily in confinement for several days when kept in water; and under these circumstances their different habits may be conveniently observed. Occasionally it will be found that they become languid and inert, especially if they have been confined a long time, but they may generally be roused into activity by

* *Cycladæ* of Gray's *Manual*, p. 277.

† See my "Monograph on the British species of Cyclas and Pisidium," *Camb. Phil. Trans.* vol. iv. p. 289.

‡ Probably in all cases in which species are really distinct, except, perhaps, in the lowest and most imperfectly organized animals, there exist some corresponding differences of habit, as well as of structure, to mark the distinction. These differences cannot be too closely inquired into by the naturalist. Whenever it is in his power to watch the habits of any supposed new species, either in its native haunts, or by keeping it in confinement, let him not pronounce positively respecting it, till he has called to his aid the valuable assistance to be derived from this source. The importance of noting the smallest fact in respect of the habits of animals has been already spoken of in the Introduction to this work (10).

the sudden application of cold spring-water; and this is by far the best method of getting a sight of the siphonal tubes which in some cases afford good distinguishing characters. Certain species, however, are at all times much more active than others. Thus the *Pisidium obtusale* * is almost always in motion, residing less at the bottom of the vessel in which it is kept than any other species in the family : it transports itself rapidly along from place to place, and appears to delight in floating masses of confervæ and other weeds.—The *P. amnicum*,† on the contrary, keeps wholly at the bottom, with the anterior half of its shell buried in the mud.—The *P. pusillum* ‡ appears to be somewhat amphibious in its habits : it frequently leaves the bottom of the vessel, and, ascending the sides, takes up its residence immediately above the edge of the water, with its shell wholly exposed : there it will remain tranquil for a length of time. The most remarkable habit, however, possessed by some species of these bivalves is that of walking, if we may so express it, on the under side of the surface of the water. This habit is common to many gasteropoda, but I am not aware of its having been observed in any of the bivalve mollusks. The action consists in the animal extending its foot along the surface, with its shell immersed, and in an inverted position. In this manner it contrives to traverse the water as though it were crawling along a solid plane. This habit is not exercised, except by

* *Monograph,* p. 301, tab. xx. f. 1—3.
† *Id.* p. 309, tab. xix. f. 2.   ‡ *Id.* p. 302, tab. xx. f. 4—6.

a few species : the *Cyclas calyculata β**, and the *Pisidium obtusale*, shew it in perfection ; whereas I never noticed it in the *P. pusillum*, or the *P. pulchellum*,† though both these will freely ascend the sides of the vessel, and the latter especially is at times very active and lively in other respects.‡

* *Monograph*, p. 298, tab. xix. f. 1.

† *Id.* p. 306, tab. xxi. f. 2—3.

‡ The above remarks on the habits of these bivalves are extracted from my Monograph ; at the time of publishing which, I was not aware of any one who had attempted to explain this singular property, possessed by certain mollusks, of walking in an inverted position immediately beneath or against the surface of the water. Since then I have met with a passage in the *Annales des Sciences Naturelles,* (2nd series, tom. xix. pp. 309—10,) in which it is alluded to by M. Quatrefages, who explains it by referring it to ciliary action. As his remarks are interesting, and enter more into the details of the phenomenon than what are given above, I will here translate the passage in question, though perhaps the phenomenon may require some further explanation than that which he offers. The passage occurs in a memoir on the *Eolidina paradoxum*, a new genus of *Nudibranchiate mollusks*, in which he finds vibratory cilia existing on every part of the surface of the body. In reference to their supposed occurrence in other *gasteropoda*, he says :—

" The peculiarity that I have just noticed, seems to me to serve to account for a fact, which has at different times attracted the attention of malacologists, and of which I have nowhere seen any very satisfactory explanation. It is known that almost all the gasteropods can walk, as it were, on the surface of a liquid, turning themselves upon their back. In this position, the lower surface of the foot coincides nearly with the superficial stratum of the liquid, and it is seen contracting itself and bending itself in different directions, as if it took its resting-place upon the stratum of air which it just grazes : the animal advances, though very slowly, without employing any other apparent means of locomo-

All the species of this family of bivalves breed readily in confinement, during the spring and summer months. They are probably ovoviviparous; and the young appear to remain for a certain period within the folds of the branchiæ previous to their exclusion, since many may be found of different sizes within the parent at one and the same time. They have the faculty of producing long before they are

tion. This phenomenon offers two distinct peculiarities, both which stand in need of explanation. First, how can an animal, specifically heavier than the liquid in which it is immersed, remain thus suspended without falling to the bottom? — In the second place, where does it find a foundation of sufficient support to enable it to move? is it in the liquid itself? or is it against the stratum of air, as some naturalists seem to allow? This last hypothesis seems to me scarcely tenable, looking to the difference of the density of air and water; from which it would follow that the resistance to be overcome in order to advance, would be incomparably greater than the effort produced by resting itself against the air. Moreover, the body of the animal possesses no appendage proper for producing this effort. The little rounded inequalities, or undulations, which the foot forms in contracting itself, cannot serve for this purpose. The same reason appears to me to be an objection to that view which regards the foot as the immediate agent in locomotion, by taking its support or resting-place in the ambient water. In that case, though the undulations of the foot might have a certain hold upon this liquid, the movement they produce is very far from active enough to create an impulse capable of overcoming the resistance presented by the water to the bulk of the body.

"But if the entire body were covered with vibratory cilia, there would be no longer anything surprising in seeing the animal use a means of locomotion which serves for the large *Planariæ*, and the *Nemertes*. This fact enables us also to clear up the first of the difficulties which we raised. Undoubtedly the pulmoni-

arrived at their full growth; and even some individuals which are themselves so immature as to possess hardly any of the distinguishing characters of the species to which they belong, frequently contain young of a sufficient size to be seen from without through the transparent valves.

ferous gasteropods can sustain themselves on the surface of the liquid by taking in a larger quantity of air. But this explanation in no way applies to those gasteropods which breathe by gills, and still less perhaps to those which, like the Actæon for example, possess no appendage ; and all these mollusks exhibit the same phenomenon. I have even seen a small Actæon move itself in exactly the same manner in the very body of the liquid. These facts find a very simple explanation in the existence of vibratory cilia; and this mode of locomotion is not more extraordinary in the case of these mollusks than it is in that of the *Infusoria,* that of the *Planariæ,* the *Nemertes,* &c."

There is no allusion, by M. Quatrefages, to the circumstance of any of the acephalous mollusks possessing this property, as well as the gasteropods.

A

CALENDAR OF PERIODIC PHENOMENA

IN

NATURAL HISTORY;

WITH REMARKS ON THE IMPORTANCE OF SUCH REGISTERS.

———

Calendaria Floræ quotannis conficienda sunt in quavis Provincia, secundum frondescentiam, efflorescentiam, fructescentiam, defoliationem, observato simul climate, ut inde constet diversitas regionum inter se.—Linn. *Phil. Bot.*

Les phases de l'existence du moindre puceron, du plus chétif insecte, sont liées aux phases de l'existence de la plante qui le nourrit; cette plante elle-même, dans son développement successif, est en quelque sorte le produit de toutes les modifications antérieures du sol et de l'atmosphère. Ce serait une étude bien intéressante que celle qui embrasserait à la fois tous les phénomènes périodiques, soit *diurnes* soit *annuels ;* elle formerait, à elle seule, une science aussi étendue qu'instructive.— Quetelet.

REMARKS

ON THE IMPORTANCE OF REGISTERS OF PERIODIC
PHENOMENA IN NATURAL HISTORY.

————

(1.) IT has been already stated in the Introduction
to this work, that a Calendar of periodic phenomena
in Natural History may be constructed for other purposes than those immediately connected with that
particular science. Observations of such phenomena may be combined with others in Meteorology,
and tend to enlarge our knowledge of particular climates. Or, without being so combined, they may
serve of themselves to point out many climatological
considerations, of the greatest importance in certain
branches of human industry dependent upon the differences which one climate exhibits compared with
another. This results from the close connection
which subsists between the phenomena of climate
and the phenomena of the animal and vegetable
worlds. The station, or habitat, together with the
geographic range, both of plants and animals, and
almost every feature in their history that is of a periodic nature, is more or less regulated by the laws

of climate, combined with the varying influence of different seasons. It has been observed that the several stages of existence in the tiny plant-aphis, or the most insignificant insect, are connected with the stages of existence in the plant which nourishes it; and this plant itself, in its gradual development, is in some measure the product of all the previous modifications of the ground and atmosphere.* It becomes therefore a most interesting study, and almost forms a science in itself, to trace out this connection, and to observe the periodic time, as well as the simultaneity, whenever it occurs, of all the resulting phenomena, considered relatively with each other.

(2.) Of the phenomena themselves, or the kinds of facts which, in reference to these points, should be attended to by the observer, we shall speak in detail presently. Some of them have been already alluded to in a former part of this work.† Such persons as have made it their occupation or amusement to collect and register these phenomena, have seldom failed to notice in many cases a coincidence of date, which is both interesting and instructive. It has been found, that, however much the seasons may differ in different years, the phenomena generally follow one another in the same order. This indeed is what we might expect, from the circumstance of any interruption in the time of their occurrence, due to seasonal

---

* *Instructions pour l'observation des phénomènes périodiques, par Quetelet.* From this memoir some of the remarks that follow are borrowed.

† *Introd.* (11.)

influence, necessarily affecting them all equally. And it follows that those which occur together any one year, will occur at or nearly the same time every other. It would not be difficult to make out a long list of such coincidences, the phenomena in which they appear being either instances from the vegetable kingdom alone, or from the vegetable and animal kingdoms together. Thus it has been noticed that the biting stonecrop (*Sedum acre*), the vine (*Vitis vinifera*), and wheat (*Triticum hybernum*), are all usually coincident in their time of flowering, which is about midsummer. A similar coincidence of date takes place most years in the flowering of the catmint (*Nepeta cataria*) and the purple loosestrife (*Lythrum salicaria*), about a fortnight after Midsummer.

Earlier in the season, we find the box (*Buxus sempervirens*), and the ground-ivy (*Nepeta glechoma*), opening their flowers most punctually together on the 3rd of April, or thereabouts.* The bugle (*Ajuga reptans*), and the horse-chestnut (*Æsculus hippocastanum*), generally flower together, in like manner, about the 3rd of May.

As examples of coincidences between the periodic phenomena of plants and those of animals, we may mention Mr. Selby's remark,† that the black-cap and willow-warbler never arrive in the north of

* I find these plants flowering together on the 3rd of April, on an average of nine years in the case of the box, and of eleven in that of the ground-ivy; and it is remarkable that the same date stands equally against both of them in the *Naturalist's Calendar* by White.

† Quoted by Mr. Yarrell in his *History of British Birds.*

England till the larch-trees are visibly green; and
that the garden-warbler and wood-warbler make
their appearance at the same time that the elm
and the oak are bursting into leaf. Indeed, the
very name imposed upon some plants and animals
is indicative of such sort of coincidences. Thus the
flos-cuculi or cuckoo-flower of the older botanists
was so called from its opening its flowers about
the time of the cuckoo's commencing his call.
Again, the wheatear, Ray tell us,* had this name
originally given to it by the Sussex people, on
account of its resorting to the Downs of that
county in such numbers, previous to its departure, just when the corn is ripe for harvest.

(3.) But, besides the above class of coincidences,
there are others which, after several years' observations, may be determined with tolerable correctness; and which, being once fixed in any given
country, may not only be instructive, but useful
in an economic point of view. We allude to those
which indicate an accordance between particular
phenomena and a certain condition of soil, or seasonal influence, fit for the varied operations of
the gardener and the husbandman. As this is a
subject which has been often treated of by former
writers, it will not be necessary to dwell long
upon it here.† We may state, however, the

---

* *Synops. Meth. Avium*, p. 75.

† See especially Harald Barck's Dissertation "On the Foliation
of Trees" (*Amœn. Acad.* tom. iii. p. 363, and *Stillingfleet's
Tracts*, p. 133); also the Preface to Stillingfleet's *Calendar of
Flora*.

grounds upon which it rests its claims to our consideration. It is obvious that to insure success in many field operations, as for instance the raising of crops, the farmer must avail himself of a particular period of year, most favourable, in respect of the state of the atmosphere and soil, for sowing the seed. Now, the determination of this period might probably be made to rest more safely on the first occurrence of some phenomenon in the vegetable world, such as the leafing of some tree or the flowering of some plant, than on the return of a fixed day, even if the weather just at that time were to be such as might be deemed favourable for the undertaking in question. Thus, the middle of March may be, in the long run, the most suitable time for sowing several kinds of grain; but we know that in some years everything at that time is locked up in frost and snow; and even if it were not so, and the weather were fine and open at the usual period of putting the seed into the ground, still its germination, and the casualties to which the young shoot is liable to be exposed after it is evolved, must depend in a great measure upon the actual condition of the soil in respect of temperature and moisture; which last must depend upon the whole character of the previous weather from the very commencement of the year. Whatever be the weather just when the farmer sows, it must make a wide difference whether the quantity of rain for several weeks previous has been in excess or not, or whether the mean temperature of the soil has reached a certain degree. Now, if he only go by the almanack in fixing his time, or if he only avail himself of the first fine week after the usual time is gone by, nothing of this sort will be taken into account: but if he know that the same conditions of soil and atmosphere, which are requisite for his purpose, are also requisite for bringing into flower or leaf any particular plant, he cannot be far out in his reckoning, if he wait for the first appearance of such plant to guide him in his operations.

(4.) It is not improbable that some natural phenomena might be found, in every month in the year, calculated to assist in indicating the fittest time for undertaking all the principal operations both in the field and garden. Of course the selection of them must be made with great care, and, as we before said, should be the result of many years' patient and accurate observation. It would be necessary with this view to take down the exact date, over a long period of time, at which such operations are now carried on; together with that of the first occurrence of the most obvious, and at the same time most characteristic phenomena, in regard of the indication they afford of certain states of weather or advances of season. By afterwards examining the results of those operations, and noting the time at which the most successful appear to have been conducted, it may be found, on looking back, what natural phenomena first showed themselves just at that period, or were coincident with the time selected. Something has been already attempted in this way,* but it would require the cooperation of many observers to arrive at any useful or trustworthy conclusions. Nevertheless, the subject is worth attending to; and if a natural calendar, constructed, as some have proposed, after this manner, did not entirely supersede the artificial one at present in use, it might at least have a subsidiary value in the case of some matters in husbandry and gardening, calling for a more than ordinary regard to the conditions of the soil, and the influence of particular seasons.

(5.) We come now to speak of another application of the Naturalist's Calendar of Periodic Phenomena, and one calculated to advance the ends of science in a more general manner. This is its application to the subject of the laws of climate, and at the same time to many important questions in animal and vegetable physiology. In reference to these ends, it is requisite that such calendars be kept in different places, upon a given plan, by as many different observers as may be prevailed upon to join in the inquiry.† When so conducted, as they have been lately, according to a system to which we shall have occasion presently to allude further, the results

* See Bark's tract, already alluded to, and Berger's *Calendarium Floræ*, in the *Amœnitates Academicæ ;* both translated by Stillingfleet, who has appended to the latter his own calendar.

† Such calendars should be combined with observations in Meteorology, according to the suggestion of Linnæus placed at the head of these remarks; *Calendaria Floræ quotannis conficienda sunt in quavis Provincia, secundum frondescentiam, efflorescentiam, fructescentiam, defoliationem, observato simul climate, ut inde constet diversitas regionum inter se.* Philosoph. Botan. p. 276.

Q

may be of great interest as well as value. By the help of such observations, made simultaneously upon a fixed number of species of plants and animals judiciously selected, we are enabled to distinguish those localities which are similarly circumstanced in respect of the phenomena to which the observations relate; and when put in possession of many such localities, we can, by drawing lines through them, indicate in a very expressive and instructive manner, the *isochronism* of these phenomena, which must be dependent upon certain conditions of climate connected with physiological conditions in the phenomena themselves. For instance, the first flowering of particular plants occurs, we know, at different times in different places; but there are some places in which, in the case of a given species, it occurs at the same time: lines passing through such places may be called *isanthesic* lines. In like manner we may have lines of equal leafing, or lines passing through those places on which particular trees come into leaf simultaneously; also lines of isochronic fructification. All these lines would be analogous to the isothermal and other lines adopted by Humboldt to denote places having the same peculiarities of temperature; and from the close connexion, before alluded to (1.), between the phenomena of climate and the phenomena of the vegetable kingdom, it would be extremely interesting to trace the relation which they bear to these last lines as well as to each other. We might ask whether the *isanthesic* lines preserved any parallelism with the *isothermal** lines, or whether in certain cases with the *isotheral*† or *isocheimal*?‡ Again: as we may conceive lines upon which a given plant comes into flower at the same time, we might have other lines upon which its flowering is retarded or advanced ten, twenty, or more days; would these lines be equidistant from each other? Also, would the lines of synchronic leafing, flowering, and fructification, for the same plant, be generally coincident?§ A number of such questions suggest themselves to the thoughtful observer. And there is no reason why we might not extend the same mode of illustration to periodic phenomena in the animal world. All those points in their history which are dependent upon time and place; such as their periods of migration and winter sleep, or of nidification and breeding; or, in the case of the insect world, their periods of metamorphosis; as well as accidental occurrences, coincident with particular seasons, such as the prevalence of certain epidemic diseases, or the unusual abundance of any particular species; these questions might receive great light from a simultaneity of observations, made with precision in many different localities, and indicated in this manner. We may allude to synchronic lines in respect of the first appearance of birds of passage, indicating their gradual

* Places having the same *mean* temperature.
† Places having the same *summer* temperature.
‡ Places having the same *winter* temperature
§ Quetelet.

progress as the seasons advance, as those which
would have an especial interest, and which would
perhaps be the simplest to determine.

(6.) Having made these remarks on the uses to
which we may apply a calendar of the principal
periodic phenomena in the vegetable and animal
kingdoms, we may go on to state, in a more methodi-
cal way, what sort of phenomena it may be desirable
to record, as well as to mention certain considerations which should be attended to by the observer.

(7.) The naturalist who collects such dates simply
for his own use, and in order to perfect as far as
possible the history of that particular class of animals or plants which is the object of his studies, or
such history as connected with the particular district in which he resides, is differently circumstanced
from him who has more general purposes in view.
Such an observer is independent of other observers.
He may make his own selection of the facts he is to
notice, or rather he will try to get together all the
facts of a periodic nature to which he can have
access. It may, nevertheless, assist him to have the
principal of these phenomena classed to his hand
under their several heads, in order that he may not
overlook any, and that his observations also may
assume more of a systematic character. With this
view, we have drawn up the following tabular
arrangement of such periodic phenomena in the
animal and vegetable kingdoms as appear most
worthy of record.

MAMMALS.

With respect to these should be noted—

The times of first appearance and withdrawal of those species which, like the bat tribe, conceal themselves during winter.

„   of commencing and waking from winter sleep, in the dormouse, badger, &c.

„   of assuming spring and winter dress, in the case of the *Mustelidæ*.

„   of coupling and parturition.

„   of shedding the antlers, in the stag and deer.

BIRDS.

In respect of migration.
{
Times of first appearance and disappearance* of regular summer and winter migrants.

„   of double passage, in those species which occur twice in the year.

„   of appearance of rare or accidental visitants.
}

In respect of song.
{
„   of commencing song.

„   of full song ceasing.

„   of re-assuming song, in those species which are heard again in the autumn.
}

In respect of breeding.
{
„   of pairing and building.

„   of laying, and commencing incubation.

„   of hatching.

„   of fledged young quitting the nest.
}

The times of moulting, whether single or double.

„   of collecting in flocks, and dispersing of the same in spring.

---

* The exact period of departure is with difficulty ascertained in the case of many species; but it should be especially attended to in that of the swallow tribe, in which it may be readily observed.

REPTILES.

The times of hybernation and first appearance in the spring.

    „   of coupling ;

    „   of laying and hatching ;   } in the serpent

    „   of casting the slough ;   tribe.

    „   of parturition, in the case of the viviparous lizard.

    „   of commencing the spring croak ;

    „   of coupling and spawning ;

    „   of the hatching of the tadpoles ;   in the frog

    „   of the tadpoles acquiring feet ;   and toad tribes.

    „   of the young frogs and toads coming on land ;

    „   of coupling and spawning ;

    „   of hatching ;   in the newts.

    „   of larvæ losing their branchiæ, and acquiring the perfect form ;

FISH.

Time of ascending rivers to spawn, in those species which quit the sea for this purpose.

    „   of leaving deep water, and approaching the shore ; in those species which never quit the sea, but appear periodically on the coast.

    „   of spawning, and hatching of the young fry.

INSECTS.

Time of first appearance.

    „   of depositing their eggs.

    „   of hatching of the young larvæ or caterpillars.

    „   of the larvæ passing into the pupa state.

    „   of the appearance of the imago or perfect insect.

    „   of hybernating, in those species that live through the winter.

    „   of the accidental occurrence of any species, either in the larva or perfect state, in unusual quantities.

MOLLUSKS.

Time of land and freshwater species quitting their respective
      winter retreats.
   „   of coupling, and laying their eggs.
   „   of hatching of the young.

The above heads of periodic phenomena relate to
the animal kingdom. In the vegetable kingdom the
observer should notice—

With respect to Phanerogamous Plants.

Time of the sap's rising in trees, indicated by the swelling of the
      buds.*

	Time when the first leaves open.
	„ when in full leaf.
Times of leaf-	„ when second, or midsummer, shoots appear.†
ing, &c.,	„ when leaves begin to change.
comprising	„ when change becomes general.
	„ when leaves begin to fall.
	„ when trees stript of leaves.

---

   * An observer has suggested, that, with respect to buds, it
would be very interesting to note, at least in the case of some of
the principal trees, whether at the end of the season they were
much or little developed. With this view, we might give the size
of them towards the end of October, by measuring their longitu-
dinal and transverse diameters. He attaches much importance to
such observations, from the circumstance that the rapidity with
which the leafing takes place the following spring, and even the
time of leafing, do not depend so much on the temperature of the
spring, as on the degree of development to which the buds had
arrived previously to winter. *Phénomènes périodiques du Règne
Végétal.* (Extrait du tom. ix. No. ii. des Bulletins de l'Acad. Roy.
de Bruxelles.)

   † The same observer alluded to in the last note thinks that the

Time of the germination of seeds, annuals more especially, noting
    also the period when sown.
    „ of herbaceous plants first appearing above ground.

Time of flow- ⎧ Time of flower-buds appearing.
ering, &c.   ⎨ „ when flowers first open.
comprising   ⎩ „ when in full flower.
                „ when going off flower.
                „ when out of flower.
Time of seed or fruit ripening.*

With respect to Cryptogamous Plants.

Time when mosses first shew their thecæ.
    „ when they ripen ditto.
    „ when confervæ, or other algæ, abound in ponds and ditches.
    „ when fungi appear, and whether in fairy rings.

If the observer extend his inquiries to field husbandry, it would be important to note—

Times of sowing different crops.
    „ of young crops appearing above ground.
    „ of haymaking.
    „ of corn, especially wheat and barley, coming into ear.
    „ of corn ripening, and commencement of harvest.
    „ of corn all carried, and fields cleared.

---

period of the second leafing, which is very noticeable in some
trees, and takes place about midsummer, is interesting, from
its marking the commencement of summer and of the greatest
heat; he considers also that it is not unlikely to be coincident
with many periodic phenomena in the animal kingdom, especially
with the appearance and hatching of certain insects.

* In the instance of the Lime, the period of the ripe seed falling is a very marked epoch, from the circumstance of the large
size of the attached bracts, which fall with it. This phenomenon,
which has been alluded to in a former part of this work (*Int.*
p. 32), should be particularly attended to.

To which may be added—

Time of occurrence of any accidental disease or blight affecting
any particular crop.

(8.) Such are some of the most important periodic
phenomena in natural history, which the attentive
observer will do well to record. It will be obvious,
on inspecting the above table, that there are a few of
these, which, in respect of *relative* date, will be the
same or nearly the same every year, and can scarcely
be called so much dependent upon season, as upon
the date of occurrence of some anterior phenomenon with which they are connected, this last alone
being influenced by the cause just alluded to. Thus
the exact time of a bird's commencing to lay may
vary with the season; but, this being determined,
the periods of incubation, and of the young being
hatched and fledged, probably follow regularly in all
cases after a given interval. Parturition in like manner in mammals is determined by the time of
coupling. Also, occasionally, we may have a variation of date, usually dependent on season, but in
certain instances varying from other causes. Thus,
cock birds which have lost their mate will sometimes protract their song considerably beyond the
usual time of ceasing. But, in general, the great
bulk of such periodic phenomena are directly connected with season and climate, and have a range
of variation according as these last vary. And
this will be particularly the case with plants, from
their being attached to the soil. With these there
is scarcely one epoch, throughout their several suc-

Q 5

cessive stages of development, which can be called fixed, even in reference to epochs that have already transpired: there is no fixed interval between the times of their coming into leaf and flower, or between the time of flowering and the time of maturation of the fruit.

(9.) The average range of variation in respect of different periodic phenomena of the above kinds appears to be about a month. By average range we mean an average deduced from the sum of the extreme ranges in the several individual cases. But of course this will greatly differ in different localities. It is also observable, though just what might be expected, that the range of variation is much greater in respect of those phenomena which occur early in the year, than of those which occur later. The flowering of plants which naturally open in January or February, or the first appearance of animals which are torpid during the winter, may be accelerated or retarded two months or more by an unusually forward or backward spring : whereas such phenomena as generally shew themselves at or about Midsummer rarely vary, in the times of their occurrence, to anything like that extent. The consequence is, as observed by Lord Bacon in the case of flowers,* that a long winter makes the earlier and later phenomena come together. The phenomena themselves, however much they vary in their respective dates, generally, as has been before observed (2.), follow one another in the same order.

* *Nat. Hist.* (edit. 7,) p. 119.

But then, to insure the correctness of this statement, we must make our observations wherever we can, as in the instance of plants we readily may do, upon the same individuals. For though all are equally subject to seasonal influence, there is a difference in individual temperament, which causes some to be much sooner impressed by this influence, and to have their development proportionably accelerated.—To this subject, however, we shall have occasion to revert presently.

(10.) Taking into consideration all that has been above stated, it appears that the naturalist or the botanist who wishes to determine the date of occurrence of any natural phenomena connected with his particular studies, even in reference to a given locality, must extend his observations over a considerable number of years to obtain the *mean time,* which is what he should always inquire after. It should be remembered that there is a mean time for the return of every phenomenon which is regulated by time at all, however great may be the departure from it in particular instances. And the observer's object should be to endeavour to find this mean, by repeating his observations through such a number of successive seasons as may insure comprising within the period seasons of as varying a character as ever occur in that particular climate. When the mean is ascertained,* it will be interest-

---

* It must be borne in mind that the *true mean* of several dates is not the mean of the *extremes,* but the mean of *all* the dates, ascertained by adding the whole together, and dividing by the exact number to which they amount.

ing to add the extreme dates in respect of early and late occurrence, and the included range in the case of each phenomenon.

(11.) We have been considering the case of the naturalist who collects these dates solely for natural history purposes, and who gets together as many as he can of every different kind. He, however, who has it in view to deduce results serving to illustrate any particular climate, and who makes observations of this nature in connexion with, as they always ought to be in connexion with, others in Meteorology, will find it better, instead of observing indiscriminately all that offers itself, to make a *selection of phenomena* in the first instance to be particularly noted, being those which are most obvious in themselves, as well as most dependent upon atmospheric causes for the time of their occurrence.—Also, if he desire to compare the results so obtained with similar results obtained in other places, it can only be by acting in union with other observers, all agreeing to observe the same periodic facts, either in some given year, with reference to climate only, or for a given number of years consecutively, with reference to climate and season combined.

(12.) Persons desirous of beginning to observe periodic phenomena might of course select their own course in this matter, and in reference to whatever ends they please. But hitherto naturalists have been unable to do more than to collect observations here and there for their own immediate use, from the want of other observers with whom to

co-operate upon some common plan.   Hence the
application of this subject to climatological purposes
has been scarcely practicable.*   The case, however, is much altered now, from the circumstance
of a number of scientific persons having agreed
to undertake a regular system of observations in
Meteorology and Natural History, with an especial view to those important questions connected
with climate, and animal and vegetable physiology, before spoken of (5).   In the paragraph
just referred to, we have already alluded to this
scheme of systematic observing, which originated,

* We would not have it inferred by this, that some valuable
attempts have not been occasionally made to combine the observations (upon the periods of vegetation principally) collected by
different observers in different places, so as to shew their connexion with, and to illustrate, the laws of climate. All we mean is,
that such attempts could hardly be productive of results equal to
what may be expected from carrying out the plan to which reference is about to be made, and which we believe is quite new as
regards European observers, though we are not sure whether a
similar one has not for some years back been acted upon in the
United States.

As a specimen of what has been effected hitherto in this way
in modern times, we may allude to an instructive paper by Mr.
Hogg, "On the Influence of the climate of Naples upon the
periods of Vegetation, as compared with that of some other
places in Europe." (*Phil. Mag. and Journ. of Sci.* 1834, vol. iv.
p. 274, and vol. v. pp. 46 and 102.) The greater part of this paper,
the author tells us, is translated from an "Essay on the Physical
and Botanical Geography of the Kingdom of Naples," by Signor
Tenore: the remainder consists of some comparative remarks on
the vegetation of England, principally compiled from White's
Naturalist's Calendar, with observations of his own.

we believe, with M. Quetelet of Brussels; and
this gentleman has been at the pains to print
and circulate ample directions for all those who
are willing to join in it, so as to insure that unifor-
mity of plan which is necessary for its success.
For our part, we consider the scheme so desirable,
and the ends which it is likely to answer so important, that we strongly recommend all willing
observers, who have not already adopted any other,
to fall in with it. They may thus lend their assistance in a common cause, and do more perhaps
for the interests of science than they could do in
any other manner. We mean not to say that observations according to their own views, and un-
combined with those of others, will be useless.
On the contrary, they must always have a certain
value, if correctly made, and made in order to
determine the *mean time* of any periodic phenomenon whatsoever. For it would be desirable to
know the mean time of all the periodic phenomena in nature without exception, as well as in
all places. But it must be immediately obvious,
from the boundless extent of such a field, what
an enormous mass of materials must be collected
in reference to such a purpose; and how hopeless would be the task of deducing any important
generalizations, for an almost indefinite time to
come, if each observer were left to choose for himself to what particular phenomena he should direct
his attention.—It is far preferable, therefore, in
the present infancy of the attempt to arrive at
any fixed laws of climate in this manner, that we

restrict our observations to such given points, as may admit of comparison in their results, and remove the risk of observers labouring to no purpose, because not labouring upon any common plan.

(13.) The meteorological observations which M. Quetelet has proposed should be made, are foreign to the subject of this work. But it may be desirable to say more on that part of his scheme which relates to the observing of periodic phenomena in Natural History.—We may state that, for the guidance of those who may be prevailed upon to join in these inquiries, and to insure a *simultaneity* both in their observations and their method of observing, he has published lists of such species of plants and animals as he and others have agreed upon to notice, (and to which it is recommended observers should exclusively confine their attention,) as well as laid down certain conditions, which should be strictly kept to, in order to give value to the results.

(14.) The principal heads of observation are similar to some of those we have already presented to the reader in a tabular form (7). They comprise the times of foliation and defoliation of upwards of a hundred species of trees and shrubs; the times of flowering and ripening of the fruit in the same species, and in about two hundred plants besides; — the first appearance and retreat of certain hybernating mammals; the times of migration of about forty different species of birds, as well as a few other points in their history of a periodic nature; the times of appearance and breeding in some of the commoner reptiles; the times of certain fish ascending rivers, or approaching the coast, to spawn; and the time of the first appearance of a limited number of insects amounting to about twenty. These lists * have been selected with much care, and with due reference to some especial points, — which deserve the consideration of observers generally in this department of scientific inquiry. Thus, in the case of the plants, it has been thought desirable to exclude all annuals and biennials, from the circumstance of their time of flowering depending in a great measure on the time when the seed was sown, so that the results would not admit of fair comparison: the only exceptions made are those of the autumnal cerealia, such as winter rye, wheat,† and barley, from their great importance and extensive cultivation; but it would be proper to notice also the time when these are sown, as well as that of their first coming into ear. There are also other grounds on which it becomes necessary to reject some plants: we must reject those, for instance, which, like the dandelion and chickweed, flower nearly the whole

* We do not think it necessary to insert the lists here, as they will be found in the last volume of the *Reports of the British Association*, (Cambridge Meeting, 1845,) according to the recommendation of a Committee appointed by the Association to consider this subject.

† The particular variety noticed in this case should be named; as some kinds of wheat, the Talavera, for instance, flower a fortnight before others.

year round ;—or which yield varieties by cultivation, varying much in their time of flowering, so
as to render it necessary to observe always the same
variety, which is often impossible ;—or which have
other species so nearly allied to them, as to make
it difficult for observers in general to distinguish
between them;—or which, lastly, are of such a kind
that the exact period of the flowers opening does
not admit of being determined with precision ; such
are the *Illecebrum*, *Aquilegia*, and others.

(15.) With due regard to the above points, the
selected species are for the most part perennial or
woody ; these last deserving especial notice from
their serving also for observations in respect of leafing. The list comprises plants from all the European
families, preference being given to those which are
most common, or which are most widely distributed,
and which have large-sized flowers ; they are also so
chosen as to offer species, some of which flower every
month in the year. There are some precautions to
be observed with respect to the particular individuals
on which our observations are made. Thus, we
should avoid those plants which have been planted
within a year previous ;—as transplanting in the
spring leads to some uncertainty in the time of leafing and flowering the season following. We should
also, for reasons already given (9.), make our observations, as much as possible, on the same individuals
every year ; which should be selected according to
the circumstances under which they grow. On
these accounts it is judged preferable to observe cultivated plants, in most instances, rather than wild

ones : we can then be sure of observing the same in two successive seasons, and we can choose those which grow in spots that are well ventilated, where they are neither too much sheltered, nor exposed to the unnatural warmth of a south wall. With regard to trees, such should be selected as grow in the open fields, not in woods, where they are very unequally screened from atmospheric influence.

(16.) The directions for observers, issued by M. Quetelet, give rules also for judging of the exact epoch of leafing, flowering, &c., in any particular case. Thus, in regard to leafing, we should choose the moment when the first leaves are so far expanded as to bring their upper surface into contact with the atmosphere, from which time the commencement of their vital functions is dated. The flowering is determined by the petals or corolla being sufficiently open to shew the anthers, which will apply equally well to the flowers of the family of the *Compositæ*, as to other cases. The fructification is regulated by the dehiscence of the pericarp, at least in dehiscent fruits, which are the great majority : in indehiscent fruits, it must be determined by their being manifestly ripe, and arrived at maturity. Lastly, the time of defoliation should be considered as come, when the greater part of the leaves of the year are fallen : it being well understood that what concerns the leaves applies only to the woody plants ; moreover excluding evergreens, in which the defoliation takes place at successive periods.*

* In the above remarks upon M. Quetelet's plan, we have not made any allusion to his proposed horary observations on the

(17.) Similar considerations, in some measure, have determined the selection of the particular species of animals recommended to be observed in the lists of M. Quetelet, as in the case of plants. Generally speaking, they are those which are most common and extensively distributed, or which possess particular interest on any other accounts. The most important perhaps are the birds, on account of their migrations, which offer points of comparison in different countries, especially valuable and instructive, as has been already alluded to (5). These are divided into four classes, comprising, *first*, those which pass the summer in central Europe ; *secondly*, those which are of regular double passage, spring and autumn, without stopping ; *thirdly*, those which pass the winter, or a part of the winter, in central Europe ; and *lastly*, those which are of accidental passage, and occur at uncertain intervals. In the selection of these species of birds, preference has been given to the land, rather than to the aquatic kinds, on account of their migrations being carried on with more regularity in all regions, and the exact time of them being more easily determined. The migrations of the latter would be not the less valuable, if correctly ascertained ; but in order to arrive at any general results, it would require more observers in marshy districts, or on the sea-coast, than can readily

diurnal periodicity of plants, (or noting the hour, varying with the time of year, at which certain species open their flowers,) as unconnected with our subject. A list of the species recommended for this purpose is given in his *Instructions,* &c. p. 14, (4to.)

be found in the present state of inquiry on this subject.

(18.) Having thus given a brief abstract of the scheme which has been recently set on foot by M. Quetelet for obtaining observations in different localities on periodic phenomena in Natural History upon one uniform plan,—we may now proceed to speak of the Calendar of such phenomena presented to the reader in the present work.   As the bulk of the observations there registered were made many years back, and long before the above scheme had been proposed, they have no immediate reference to it, nor will they admit of close comparison with any single observations made subsequently : indeed many of them are not much adapted to assist in questions of a climatological character.   It has been already stated, that for such purposes a *selection* of phenomena would be the most serviceable (11.); whereas the present calendar is of a more general nature, and intended rather for the use and guidance of observers of all classes.   Its main object was to endeavour to ascertain the mean time of occurrence of the various periodic phenomena most noticeable in our own immediate neighbourhood, and for the furtherance of that department of natural history in reference to a given locality.—In the present work, it is intended in some measure to take the place of White's Naturalists' Calendar.   With this view, the phenomena which White has recorded have been revised ; and some rejected, and others added, according as the subject, especially when taken in connexion with the circumstance of the increased attention of late paid to it, seemed to require; for, in reference to this last matter, it may still be of some service. Though it cannot be directly compared with the results of any one year's observations made upon the plan which it is proposed hereafter to adopt,— yet, according to what was before stated (12.), it must always have a certain value, so far as it assists in determining the *mean date* of any periodic phenomena in a given district. To fix the date of an occurrence in different localities with reference to a given season, the observations must be made simultaneously, or during the same year; but to fix it in reference to climate generally considered, and independent of the variation caused by seasons of different character, they may be made at any time, provided only they are continued a sufficient number of years to eliminate the error arising from this source. A mean date thus obtained will always be available in any questions of climatological research with which it is connected. It stands ready for use when wanted; and against the time when dates of the same phenomena, similarly determined in other places, may be marshalled along with it. Thus, for example: M. Quetelet has observed that the time of the lilac's flowering at Brussels is the 1st of May:* here then is a given fact always ready for embodying with others, when obtained, to unite in forming a synchronic line in reference to the flowering of that particular plant, or, as we have called it, an isanthesic line (5). In like manner, the mean time of any other periodic phenomenon, when once determined for a given locality,

* *Instructions*, &c. p. 2.

preserves its value, and may prove of use, if not immediately, yet at some future day, for such purposes as the above.

(19.) It is true that the mean date of the greater number of the periodic phenomena in the accompanying calendar may be little more than the most rude approximation; and, in many instances, there has been no more than a solitary entry in one year. Of course it is not to be expected that all which have been registered are of the same value in this respect: neither, as before said, is it exclusively our object to give a table of such dates in reference to climatological inquiries. Entries, which would prove of little or no use in such investigations, may still interest the naturalist. Pains, however, have been taken to make known exactly what the precise value of each recorded observation is, as regards its approach to a correct mean time, by annexing in all cases the number of years in which the observation was repeated, and from which the mean, as it stands in the calendar, was deduced.

(20.) This calendar, also, though it may not admit of comparison with those formed upon the plan proposed by M. Quetelet, may still assist observers who contemplate acting upon that plan. This it may do, by guiding them to the time when any of the selected phenomena generally occur, and so put them upon the watch, as these phenomena successively follow one another, throughout the several months of the year. Each month in a general way has its own distinguishing occurrences, and it may be convenient to have these arranged to one's hand, even if the arrangement is not in all cases in such exact accordance with nature as more extended observation would make it. Also, though the number of species (at least of plants) in M. Quetelet's list is much greater than in the present calendar, still, as this last is mainly intended for English observers, and comprises most of the species given by M. Quetelet which are found in Britain, it may be of much collateral advantage in this respect.

(21.) The grounds of selection, in respect of the phenomena registered in this calendar, are for the most part similar to those already alluded to in the case of the lists published by M. Quetelet. (15.) (17.) Such as belong to the animal kingdom relate principally to birds and insects. In the class of birds, they are generally those which have a more or less marked reference to, or connexion with, some fixed period of the year, especially such as are characteristic of the season, and easily seized upon by the observer. At the same time there are a few which, without affording any particular indications of this nature, may tend to illustrate and improve our knowledge of the species to which they severally relate.

In the case of insects, nothing is noted, with a few exceptions, but the time of their first appearance. Of course, in this class, the list of species inserted might be almost indefinitely extended. It has been thought proper, however, to confine it, for the most part, to such as are generally common and well known, or which at least occur in greater or less plenty each year in the neighbourhood in which

the observations were made; to such as are not
liable to be mistaken, or confounded with other
species closely allied to them; or to such, lastly, as
particularly shew themselves at fixed periods, and
are most regular in their appearance.

(22.) In regard of plants, nearly the same points
have been attended to, as those recommended by M.
Quetelet. — The species given in the calendar are
mostly common ones and widely dispersed in this
country; the greater number, though by no means
all, are perennials; also for the most part plants growing wild, though a few are only cultivated, the choice
of these last being regulated by the circumstance of
their being generally common in gardens, and well
adapted to marking the progress of the seasons.   It
should be stated, that, in the case of the wild species,
garden-plants have been preferred, when they were
to be had, both for the convenience of being able to
watch them more narrowly, as well as for the
security of being able to note the same individual
in successive years:—those only have been selected
for this purpose, which grew in spots moderately
exposed, and not affected in their time of flowering
by any undue influence of heat or shelter.

In addition to the observations on plants and
animals, there are a few inserted relating to field
husbandry, such as the times of sowing some of the
principal crops, and the times of haymaking and
harvest, which will not be deemed unimportant by
those observers to whom the periodic phenomena of
nature are a subject of interest and research.

(23.) For the assistance of those who are disposed to commence observations on the plan recommended by M. Quetelet, but who are not in possession of his lists, it has been thought useful in the following calendar to mark with an asterisk (*) all those periodic phenomena which are included in the number proposed by him to be attended to. Certain others are distinguished by a dagger (†), which, though not given by him, appear also to be deserving of notice, or at least by observers in this country, from their being, as it has been termed, "more than ordinarily prognostic" of the season at which they occur.

(24.) It has been endeavoured in this calendar, as before stated (18), to ascertain, so far as the observations would allow, the mean time of the occurrence of the several phenomena therein noted. At least there are many of the phenomena, in the case of which this endeavour has been more particularly made. The observations were all made in the immediate neighbourhood of Swaffham Bulbeck, for a number of years consecutively, commencing with 1820, and ending with 1831. During this period, there were some seasons as remarkably forward as others were backward; there were also some very hot and fine summers, as well as cold and wet ones.—The observations therefore extend over a period of twelve years. Those which have been repeated each year, or during any number of years short of the whole, but exceeding one, have their dates registered under the columns of *mean*, *earliest*, and *latest*, to the first of which is attached a subordinate column indicating what the exact number of years may have been,

R

from which the mean is calculated. When the observation has been made in only a single instance, the annexed date has been entered under the head of mean, as a place which it is entitled to hold till other observations of the same phenomenon point out what its exact value is in this respect. When only two observations have been made, and both these have occurred, as in a few instances, upon the same day, the entry has been registered under the same head, with the figure 2 annexed to it. When but two observations have been made, and they have occurred on two consecutive days, these two days have been bracketed together under the head of mean. Lastly, it may be observed, that all the entries under one date, to whatever class of phenomena they may belong, are bracketed together in such a manner that the eye may catch, at one glance, the several coincidences as they successively occur, without having to sort them out from the first of the three columns above spoken of.

(25.) Perhaps the most perfect portion of the calendar, in respect of the number of years in which the observations have been repeated, is that which relates to the periods, at which different birds commence song, and at which our principal summer visitants make their first appearance. And these phenomena are also amongst the most important, in respect of the indications they afford of the time of year, and the character of the season. Many of the observations, likewise, connected with the leafing of trees, and the flowering of the more common plants, are perhaps sufficiently numerous, in certain in-

stances, to fix the mean time of such periodic phenomena with tolerable precision. In others, no
doubt, the date registered would be liable to considerable alteration from observations continued over
a longer period of years.

(26.) Each entry must be considered in itself
alone, and as an approximation more or less to the
mean period of occurrence of the phenomenon recorded,—not as in necessary connection with any
other, with which, if so viewed, it may appear incon-
sistent, from the observations not being equally
extensive in the two cases. Thus, to illustrate our
meaning by a particular instance, there is of course
a mean time, with every species of bird, for laying
and hatching: if the true mean were really deduced
and known, in respect of each of these two phenomena, the interval between them would be the exact
period occupied in incubation. Looking, however,
at the present calendar, and without remembering
that each date is but an approximation to the true
one, this approximation being sometimes much
nearer than others, it may appear otherwise, and
this interval be thought much shorter than in reality
it is.

(27.) This circumstance has suggested the pro-
priety of inserting the dates of the various phenomena observed as they occurred in succession in
some one year, and without reference to obtaining a
mean time; by which arrangement, perhaps, their
true order of sequence can be more correctly traced.
And accordingly such a column in the calendar has
been annexed to the others already spoken of (24),
and set aside exclusively for the entries of a given

year, the one chosen being that which has just
elapsed, viz. 1845 : and it will have an especial interest, perhaps, compared with the results of the
series of years registered in the other columns, from
the circumstance of its having been an *extreme* year ;
and, throughout each of the seasons, as backward
as, if not more so than, any of those which are included in the above series. At the same time it
must be observed, that during this year only those
phenomena are registered which happened to be
observed,—and that therefore there are many blanks,
some perhaps most important ones, in reference to
other phenomena registered before.

(28.) It may be as well to state that the scientific names of the several species of plants and animals noted in this calendar, are believed to be in
most instances those adopted by the latest authors
in their respective departments. The names of the
birds are the same as those given by Yarrell in his
work on British Birds: the names of the mammals
and the reptiles are from Bell: the names of the
plants, (at least the British ones) from Babington's
Manual. In all cases, however, the English names
are prefixed, where there are any generally familiar
to the English reader. Many of the insects have
no English name at all; this, however, is not the
case with the lepidoptera, to which the same English ones have been applied as those given by
Haworth in his well-known work on this order.
The Latin names of the insects generally are those
of Stephens.

(29.) It only remains, in conclusion, to express a
hope that all errors which the reader may notice, or be

led to suspect, in this calendar, will be considerately pardoned. The bulk of the observations having been made and registered many years back, it is extremely probable that there may be some wrong entries by mistake in the journal from which they are now copied, though it is believed that they are generally correct. It will also sometimes happen that the first occurrence of a particular phenomenon is in reality previous to the day of its being first noticed; and this is especially likely to be the case with respect to the appearance of birds and insects, which cannot be watched with the same exactness as plants, and in regard of which the attention is not given to any particular individuals but to the species generally. It may be added that the calendar here offered to the public is of too comprehensive a kind to admit of being more than an approximation to a true Calendar of Nature. To form such a calendar, even for Natural History purposes alone, would require a combination of observers, each confining himself to one particular department, and in the enjoyment of such advantages, in the way of situation and leisure, as the undertaking, in order to be at all successful, demands. If the present imperfect attempt should anywise be conducive to such an end, by stimulating others to the prosecution of this subject, or should afford either interest or instruction to young observers, for whom principally this calendar has been arranged, the author will be well satisfied, as well as repaid for the trouble which it has cost him, in putting it together.

CALENDAR OF PERIODIC PHENOMENA IN NATURAL HISTORY,

AS OBSERVED IN THE NEIGHBOURHOOD OF SWAFFHAM BULBECK.

Abbreviations used :

ap. . . . . first appearance.
sg. com. . . song commences.
sg. ceas. . . song ceases.
sg. reas . . song reassumed.
l. . . . . first opening of leaves.
fl. . . . . first opening of flowers.

The asterisk (*) distinguishes those phenomena which are recommended for observation in the lists of M. Quetelet. The dagger (†) marks certain others which, though not included in those lists, are well deserving of attention as good prognostics. Italic characters are used for the scientific names : where these alone occur, there are no English names generally received.

JANUARY.

Jan.	Phenomena.	Date of occurrence ; 1830—31.					Date in 1845.
		Mean.(a)		Earliest.		Latest.	
5	Wren (*Troglodytes europæus*), sg. com.	Jan. 5	11	Jan. (b)1		Jan. 18	
8	Redbreast (*Erithaca rubecula*), sg. com.	,, 8	11	,, 1		,, 28	
11 {	Common Bunting (*Emberiza miliaria*), note com.	,, 11	5	,, 1		,, 19	
	Trichocera hiemalis ap.	,, 11	11	...		...	Feb. 18
13	Furze (*Ulex europæus*), fl.	,, 13	11	,, 1		Feb. 26	Apr. 4
15 {	Marsh Titmouse (*Parus palustris*), note com.	,, 15	8	,, 1		,, 17	
	Pale perfoliate Honeysuckle (*Lonicera caprifolium*), l.	,, 15	5	,, 1		Jan. 28	
17	Hedge Accentor (*Accentor modularis*), sg. com.	,, 17	11	,, 1		Feb. 10	
19	Cole Titmouse (*Parus ater*), note com.	,, 19	5	,, 7		,, 7	Feb. 15
21	Starlings (*Sturnus vulgaris*), resort to buildings	,, 21	3	,, 9		,, 8	
22	*Mezereon (*Daphne mezereum*), fl.	,, 22	2	,, 11		,, 2	Feb. 15

	Species					
23	Skylark (*Alauda arvensis*), sg. com.	Jan. 23	12	Jan. 13	Feb. 21	Jan. 7
24	*Great Titmouse (*Parus major*), sg. com.	,, 24	11	,, 7	,, 11	Feb. 5
25	{ Hazel (*Corylus avellana*), fl.(c)	,, 25	11	,, 1	,, 20	Jan. 23
25	{ Hepatica (*Hepatica triloba*), fl.	,, 25	9	,, 1	Mar. 25	Jan. 18
26	{ Missel Thrush (*Turdus viscivorus*), sg. com.	,, 26	12	,, 1	Feb. 28	Mar. 10
26	{ *Winter Aconite (*Eranthis hiemalis*), fl.	,, 26	10	,, 1	,, 13	Jan. 24
27	House-flies ap. on windows (d)	,, 27	7	,, 3	Mar. 4	
28	{ Stinking Hellebore (*Helleborus foetidus*), fl.	,, 28	12	,, 2	,, 3	Jan. 25
28	{ *Daisy (*Bellis perennis*), fl.	,, 28	11	,, 12		Apr. 4
29	*Hive Bee (*Apis mellifica*), comes abroad	,, 29	11	,, 1	Feb. 25	Mar. 31
30	{ *Snowdrop (*Galanthus nivalis*), fl.	,, 30	12	,, 21	,, 16	Feb. 5
30	{ Double Daisy (*Bellis perennis, fl. pl.*), fl.	,, 30	4	,, 2	Mar. 28	Apr. 13
31	{ *Song Thrush (*Turdus musicus*), sg. com.	,, 31	12	,, 1	Feb. 23	Feb. 26
31	{ *Peziza coccinea*, ap.	,, 31	2	,, 2	Mar. 1	

FEBRUARY.

	Species					
1	{ *Pied Wagtail (*Motacilla yarrellii*), first seen	Feb. 1	4	Jan. 2	Mar. 8	Feb. 19
1	{ *Chaffinch (*Fringilla cælebs*), sg. com.	,, 1	12	,, 7	Feb. 14	
	{ Flocks of Greenfinches (*Coccothraustes chloris*), separate	,, 2	2		Feb. 14	
2	{ Tawny Owl (*Syrnium aluco*), hoots	,, 2	5	Jan. 24	Feb. 14	
4	Field Speedwell (*Veronica agrestis*), fl.	,, 4	7	,, 3	Apr. 5	

(a) The small figures in the second column under the head of Mean, indicate the number of years from which the mean is deduced, when the observation has been made more than once.

(b) All those phenomena which are referred to January 1st as the earliest date, may be considered as occasionally shewing themselves in December of the previous year.

(c) This entry refers to the opening of the male catkins: the female blossoms do not usually shew themselves till a few days later.

(d) This refers to the period of their being first roused to a state of activity in a room without a fire.

FEB.	PHENOMENA.	Date of occurrence; 1820—91.			Date in 1845.
		Mean.		Earliest. Latest.	
5	*Spurge-Laurel (*Daphne laureola*), fl.	Feb. 5	12	Jan. 3 Mar. 7	Jan. 25
8	*Small smooth Eft (*Triton punctatus*), ap. in ponds	,, 8	4	Mar. 22	
9	*Gymnostomum ovatum*, ripens its capsules	,, 9		Jan. 2 Mar. 22	
11	House Pigeon (*Columba livia*, var. *domestica*), lays	,, 11			Mar. 30
11	Primrose (*Primula vulgaris*), fl.	,, 11	12	Jan. 3 Mar. 21	,, 21
12	*Blackbird (*Turdus merula*), sg. com.	,, 12	12	,, 18 ,, 7	Jan. 29
12	Butcher's-broom (*Ruscus aculeatus*), fl.	,, 12	8	,, 3 Apr. 1	Feb. 25
13	Partridge (*Perdix cinerea*), pairs	,, 13	8	,, 12 Mar. 12	
14	*Gold-crested Wren (*Regulus cristatus*), sg. com.	,, 14	10	,, 4 ,, 31	Feb. 27
14	*Elder (*Sambucus nigra*), l.	,, 14	12	,, 2 ,, 22	,, 27
15	*Yellowhammer (*Emberiza citrinella*), sg com.	,, 15	12	,, 30 ,, 4	
15	Red Dead-nettle (*Lamium purpureum*), fl.	,, 15	8	,, 2 ,, 13	June 3
16	Small bloody-nose Beetle (*Timarcha coriaria*), ap.	,, 16	3	,, 30 ,, 18	Mar. 26
18	Jackdaws (*Corvus monedula*), resort to chimneys	,, 18	5	,, 25 ,, 27	
18	*Common Honeysuckle (*Lonicera periclymenum*), l.	,, 18	6	,, 29 ,, 24	
18	Dandelion (*Taraxacum officinale*), fl.	,, 18	9	,, 1 ,, 29	
19	*Spring Crocus (*Crocus vernus*), fl.	,, 19	10	Feb. 3 ,, 3	Mar. 24
21	*Greenfinch (*Coccothraustes chloris*), sg. com.	,, 21	12	,, 1 ,, 13	,, 20
21	*Velia currens*, ap. on the surface of streams	,, 21	4	,, 9 Feb. 26	
22	*Lesser Periwinkle (*Vinca minor*), fl.	,, 22	12	Jan. 2 Apr. 25	Apr. 21
23	Earthworms lie out	,, 23	8	,, 23 Mar. 30	Mar. 29
23	*Sweet-scented Coltsfoot (*Tussilago fragrans*), fl.	,, 23	3	,, 23 ,, 20	

24	*Japan Kerria (*Kerria japonica*), l.	Feb. 24	5	Jan. 31	Mar. 12	Mar. 28
25	*Ring-dove (*Columba palumbus*), coos	" 25	12	Feb. 2	" 19	" 26
28	{ *Alder (*Alnus glutinosa*), fl.	" 28	12	Jan. 21	" 20	
	{ *Yew (*Taxus baccata*), fl	" 28	12	Feb. 8	" 28	Mar. 29

MARCH.

Mar.						
1	Heath Snail (*Helix ericetorum*), comes abroad (a)	Mar. 1	3	Jan. 31	Mar. 30	
2	*Pilewort (*Ranunculus ficaria*), fl.................	" 2	12	" 21	" 28	Mar. 11
3	{ *Rooks (*Corvus frugilegus*), build	" 3	12	Feb. 12	" 13	" 14
	{ Flocks of Wild Geese return northwards.	" 3	2	Mar. 1	" 5	
4	{ Ladybird (*Coccinella 7 punctata*), ap.	" 4	10	Jan. 3	May 18	
	{ Marsh-Marygold (*Caltha palustris*), fl	" 4	9	Feb. 9	Mar. 29	
	{ Missletoe (*Viscum album*), fl	" 4	8	" 10	" 28	
	Stock-dove (*Columba ænas*), note com.	" 5	4	" 23	" 24	Mar. 14
	Whirlwig Beetle (*Gyrinus natator*) ap	" 5	7	Jan. 25	Apr. 12	Apr. 4
5	*White Dead-nettle (*Lamium album*), fl.........	" 5	8	" 19	" 22	" 21
	Common Whitlow-grass (*Draba verna*), fl.	" 5	9	" 21	" 10	
6	{ Cushion-moss (*Grimmia pulvinata*), ripens its capsules ...	" 6	2	Feb. 13	Mar. 25	
	{ *Dor-Beetle (*Geotrupes stercorarius*), ap	" 6	6	Jan. 30	Apr. 6	
7	{ Broods of small Coleopterous Insects on wing (b)	" 7	7	Feb. 19	Mar. 30	
	{ *Sweet Violet (*Viola odorata*), fl	" 7	12	Jan. 25	" 26	Mar. 29
8	House Pigeon hatches	" 8	2	Feb. 28	" 16	

(a) I observe that this species of snail, which is everywhere common, is always much earlier in coming abroad than either the *Helix aspersa* or the *H. nemoralis*.

(b) These broods of small coleopterous insects, consisting chiefly of *Curculionidæ* and *Staphylinidæ*, are alluded to in a former part of this work, as coming on wing the first mild spring day that may occur in February or March.

MAR.	PHENOMENA.	Date of occurrence; 1820–31.				Date in 1845.
		Mean.		Earliest.	Latest.	
8	†Frog (*Rana temporaria*), croaks	Mar. 8	2			
9	{ Hooded Crow (*Corvus cornix*), last seen	,, 9	4	Feb. 12	Mar. 25	
	Gossamer floats (a)	,, 9	9	,, 26	Apr. 4	
10	{ Turkey-cock (*Meleagris gallopavo*), struts and gobbles (b)	,, 10	9	,, 20	Mar. 29	Mar. 31
	*Brimstone Butterfly (*Gonepteryx rhamni*), ap.	,, 10	12	,, 2	,, 2	,, 15
	*Apricot (*Armeniaca vulgaris*), fl.	,, 10	10	,, 20	,, 31	
11	Creeper (*Certhia familiaris*), spring note com.	,, 11	3	,, 10	Apr. 9	
	Ivy-leaved Speedwell (*Veronica hederifolia*), fl.	,, 11	7	,, 1	,, 3	
12	*Rhyphus fenestralis*, ap.	,, 12	5	Mar. 3	Mar. 26	Jun. 2
	*Gooseberry (*Ribes grossularia*), l.	,, 12	12	,, 2	,, 27	Mar. 30
13	{ *Peach (*Persica vulgaris*), fl.	,, 13	11	Feb. 20	,, 31	,, 27
	†Common Coltsfoot (*Tussilago farfara*), fl.	,, 13	12	,, 24	Apr. 1	Apr. 4
	*Daffodil (*Narcissus pseudonarcissus*), fl. (c)	,, 13	12	,, 28	Mar. 21	,, 4
14	†Peacock (*Pavo cristatus*), screams (d)	,, 14	5	,, 19	Apr. 1	
	*Red Ant (*Formica*), ap.	,, 15	8	Jan. 18	,, 19	
15	{ *Common Gnat (*Culex pipiens*), ap.	,, 15	2	Mar. 1	Mar. 30	May 3
	*Japan Kerria (*Kerria japonica*), fl.	,, 15	5	Feb. 24	Apr. 4	Apr. 21
16	Woodcock (*Scolopax rusticola*), last seen	,, 16	16	Mar. 10	Apr. 23	
	*Frog (*Rana temporaria*), spawns.	,, 16	9	,, 4	,, 25	
17	{ Pied Wagtail, spring note com.	,, 17				
	*Scarlet satin Mite (*Trombidium holosericeum*), ap. (e)	,, 17	2	Feb. 27	Apr. 4	
	*Syringa (*Philadelphus coronarius*), l.	,, 17	9	,, 24	Mar. 30	Apr. 12

17	*Aspen (*Populus tremula*), fl.	Mar. 17	5	Mar. 4	Mar. 30	Mar. 27
	Oats (*Avena sativa*), sown	,, 17	5	Feb. 22	Apr. 21	Apr. 4
18	*Lilac (*Syringa vulgaris*), l.	,, 18	12	Mar. 2	Apr. 1	
	*Dog-rose (*Rosa canina*), l.	,, 18	7	Feb. 23	,, 7	
	Hygrometric Cord-moss (*Funaria hygrometrica*), ripens its capsules	,, 18	4	Jan. 14	Apr. 14	
19	*Black Currant (*Ribes nigrum*), l.	,, 19	.	...	...	Apr. 9
	Blue Navelwort (*Omphalodes verna*), fl. (f)	:, 19	3	Feb. 26	Apr. 2	Mar. 26
	Willows open their catkins (g)	,, 19	7	Mar. 11	Mar. 27	
20	*Pipistrelle Bat (*Vespertilio pipistrellus*), comes abroad	,, 20	5	,, 3	Apr. 29	May 20
	*Humble-Bee (*Bombus*), ap.	,, 20	12	,, 4	Mar. 30	Mar. 31
	*Banded Snail (*Helix nemoralis*), comes abroad	,, 20	8	Jan. 31	Apr. 28	May 17
21	*Elater lineatus*, ap. (h)	,, 21	2	Mar. 7	,, 5	
	Weeping Willow (*Salix babylonica*), l.	,, 21	8	Feb. 18	,, 13	Apr. 23
	Sweet-briar (*Rosa rubiginosa*), l.	,, 21	6	Mar. 8	,, 3	
	*Common Elm (*Ulmus campestris*), fl.	,, 21	12	,, 7	,, 5	Apr. 4
	Barley (*Hordeum vulgare*), sown	,, 21	3	,, 15	Mar. 30	

(a) I have generally observed this phenomenon twice in the year, spring and autumn, the same as White, whose expressive term "floats" is here retained. The autumnal entry will be found further on.

(b) This entry is also made in White's own language;—the phenomenon itself indicates those feelings in the cock bird, which are connected with the approach of the breeding-season.

(c) In 1834, a remarkably forward season, this plant was in flower on the 28th of January.

(d) This does not refer to the ordinary call or cry of the peacock, but to a peculiar scream, uttered only by the male bird, when under the influence of sexual desire, and very characteristic of the first warm weather that occurs in early spring.

(e) This mite, one of the largest and most conspicuous of its tribe, may be frequently observed in gardens, crawling upon the flower-beds, in the early spring.

(f) In the present year (1846), the spring of which has been unprecedentedly forward, this plant was in flower on the 8th of January!

(g) No particular species is mentioned, as several open their catkins about the same time: the species, too, are with difficulty discriminated from each other. Yet the general phenomenon, as indicative of spring, is worth noticing.

(h) The larvæ of this species, and the *E. sputator* noticed further down, are the *wireworms* so destructive in agriculture, by attacking the roots of plants.

Mar.	Phenomena.	Date of occurrence; 1820—31.				Date in 1845.
		Mean.		Earliest.	Latest.	
22	*Magpie (_Pica caudata_), builds	Mar. 22				
	Elater sputator, ap.	,, 22				
	Badister bipustulatus, ap.	,, 22				
	Creeping Crowfoot (_Ranunculus repens_), fl. (ᵃ)	,, 22	6	Jan. 6	May 31	Apr. 16
23	*Common Linnet (_Linota cannabina_), sg. com. (ᵇ)	,, 23	12	Feb. 22	Apr. 14	,, 3
	*Oil Beetle (_Proscarabæus vulgaris_), ap.	,, 23	12	Mar. 4	,, 7	,, 16
	*Whitethorn (_Cratægus oxyacantha_), l.	,, 23	12	,, 10	,, 12	
24	_Carabus nemoralis_, ap.	,, 24	4	Feb. 27	,, 20	
	*Red Currant (_Ribes rubrum_), l.	,, 24	9	Mar. 14	,, 8	,, 12
	Wall-flower (_Cheiranthus cheiri_), fl.	,, 24	6	Jan. 9	,, 28	,, 24
25	Green Woodpecker (_Picus viridis_), cries	,, 25	5	Feb. 25	,, 17	
	Jackdaw (_Corvus monedula_), builds	,, 25	7	Mar. 7	,, 7	
	Rook (_C. frugilegus_), lays	,, 25	3	,, 17	,, 2	
	Earwig (_Forficula auricularia_), ap.	,, 25	3	Feb. 1	,, 28	
	*White Poplar (_Populus alba_), fl.	,, 25	6	Mar. 7	,, 13	
26	*_Pæcilus cupreus_, ap.	,, 26	3	Mar. 10	Apr. 6	Jun. 3
	Dog's Mercury (_Mercurialis perennis_), fl.	,, 26	5	Feb. 15	,, 10	
27	*Common Toad (_Bufo vulgaris_), spawns	,, 27	7	Mar. 16	,, 5	
	Six-cleft Plume-Moth (_Alucita hexadactyla_), ap.	,, 27	8	Feb. 3	May 5	
	*Small-tortoiseshell Butterfly (_Vanessa urticæ_), ap.	,, 27	3	Mar. 10	Apr. 15	Apr. 3
	Harpalus æneus, ap.	,, 27	7	,, 12	,, 6	
	Hyacinth (_Hyacinthus orientalis_), fl.	,, 27		,, 7	,, 9	
28	Domestic Goose hatches	,, 28				

	Phenomenon					
28	Tawny Owl (*Syrnium aluco*), lays	Mar. 28	10	Mar. 9	Apr. 12	Apr. 4
	Privet (*Ligustrum vulgare*), l.	,, 28	10	,, 12	,, 16	,, 20
	*Gooseberry (*Ribes grossularia*), fl.	,, 28	3	,, 20	,, 6	
	*Almond (*Amygdalus communis*), fl.	,, 28	9	Feb. 19	May 8	
	*Large bloody-nose Beetle (*Timarcha tenebricosa*), ap.	,, 28	8	Mar. 9	Apr. 10	
	Bramble (*Rubus fruticosus*), l.	,, 28	12	,, 12	,, 15	
29	*Hazel (*Corylus avellana*), l.	,, 29	...	Mar. 12	Apr. 15	Apr. 22
	Yellow Figwort (*Scrophularia vernalis*), fl.	,, 29	...	...	...	,, 23
	†Dwarf purple Iris (*Iris pumila*), fl.	,, 29	2	,, 16	,, 10	,, 24
30	*Bladder-nut (*Staphylea pinnata*), l.	,, 30	5	Feb. 5	,, 20	,, 19
	†Cowslip (*Primula veris*), fl.	,, 30	11	Mar. 14	,, 13	
	Grape-Hyacinth (*Muscari racemosum*), fl.	,, 30	5	,, 26	,, 15	
	Domestic Duck (*Anas boschas var. domest.*) hatches	,, 30	5	Feb. 28	,, 16	
31	*Peacock Butterfly (*Vanessa io*), ap.	,, 31	8	Mar. 13	,, 18	Apr. 4
	*Mealy-tree (*Viburnum lantana*), l.	,, 31	9	,, 15	,, 15	,, 22
	*Horse-chestnut (*Æsculus hippocastanum*), l.	,, 31	12	,, 16	,, 17	,, 21
	Hairy Violet (*Viola hirta*), fl.	,, 31	9			

APRIL.

	Phenomenon					
Apr. 1	Wych Elm (*Ulmus montana*), fl.	Apr. 1	3	Mar. 23	Apr. 10	Apr. 17
	†Solid-rooted Fumitory (*Corydalis solida*), fl.	,, 1	3	,, 27	,, 6	
	*Ivy (*Hedera helix*), berries ripe.	,, 1	4	,, 14	,, 17	
2	*Peach (*Persica vulgaris*), l.	,, 2	7	,, 18	,, 16	Apr. 16
	†Larch (*Larix europæa*), l.	,, 2	7	,, 21	,, 11	

(a) The period of first flowering in this plant, though a perennial, appears to have a great range, and is very uncertain.
(b) The commencement of song in this species is coincident with the breaking up of the winter flocks, the several individuals then betaking themselves to gardens and shrubberies.

APR.	PHENOMENA	Date of occurrence; 1820—91.				Date in 1845.
		Mean.		Earliest.	Latest.	
2	*Spring Bitter-vetch (*Orobus vernus*), fl.	Apr. 2				Apr. 23
	*Virginian Lungwort (*Pulmonaria virginica*), fl.	,, 2	7	...	Apr. 14	,, 23
	†Chiff-chaff (*Sylvia hippolais*), note first heard	,, 3	3	Mar. 15	,, 12	,, 19
	†Pheasant (*Phasianus colchicus*), utters its spring crow (a)	,, 3	3	,, 25	,, 17	,, 26
	Crab (*Pyrus malus*), l.	,, 3	10	,, 15	,, 16	,, 18
3	*Apricot (*Armeniaca vulgaris*), l.	,, 3	7	,, 20	,, 16	,, 20
	*Cherry (*Prunus cerasus*), l.	,, 3	6	,, 20	,, 14	,, 19
	*Barberry (*Berberis vulgaris*), l.	,, 3	3	,, 26	,, 23	,, 19
	*Ground-ivy (*Nepeta glechoma*), fl.	,, 3	11	,, 9	,, 22	,, 20
	*Box (*Buxus sempervirens*), fl. (b)	,, 3	9	,, 10	,, 16	,, 22
	*Plum (*Prunus domestica*), l.	,, 4	7	,, 16	,, 11	
4	Black Poplar (*Populus nigra*), fl.	,, 4	2	,, 28	,, 6	
	Lombardy Poplar (*P. dilatata*), fl.	,, 4	4	Apr. 2	,, 23	
	*Fieldfares (*Turdus pilaris*), last seen	,, 5	2	Mar. 3	,, 14	
5	Blackthorn (*Prunus spinosa*), l.	,, 5	8	,, 12	,, 17	
	*Crown-Imperial (*Fritillaria imperialis*), fl.	,, 5	10	,, 13	,, 8	Apr. 25
	Moorhen *Gallinula chloropus*), lays	,, 6	8	Apr. 2	,, 20	
6	Turnip-fly (*Haltica nemorum*), ap.	,, 6	3	Mar. 25		
	Rue-leaved Saxifrage (*Saxifraga tridactylites*), fl.	,, 6	3			
	Blackbird (*Turdus merula*), lays	,, 7	6	Mar. 16	May 1	
7	*Common Elm (*Ulmus campestris*), l.	,, 7	12	,, 1	Apr. 22	Apr. 25
	*Laburnum (*Cytisus laburnum*), l.	,, 7	3	,, 28	,, 17	,, 26
	†Blackthorn (*Prunus spinosa*), fl.	,, 7	10	,, 15	,, 20	

	Species					
7	*Plum (*Prunus domestica*), fl.	Apr. 7	9	Mar. 17	Apr. 19	Apr. 23
	*Ringed Snake (*Natrix torquata*), comes abroad	,, 8	6	,, 25	,, 28	Apr. 20
8	*Pear (*Pyrus communis*), l.	,, 8	7	,, 20	,, 17	,, 14
	*Raspberry (*Rubus idæus*), l.	,, 8	7	,, 21	,, 27	
	Larch (*Larix europæa*), fl.	,, 8	5	Apr. 1	,, 19	Apr. 23
	Missel Thrush (*Turdus viscivorus*), lays	,, 9	3	,, 2	,, 15	,, 24
	Rook (*Corvus frugilegus*), hatches	,, 9	12	,, 3	,, 17	May 14
9	*Birch (*Betula alba*), l.	,, 9	11	Mar. 23	,, 22	
	*Evergreen Alkanet (*Anchusa sempervirens*), fl.	,, 9				Apr. 20
	Turnip (*Brassica rapa*), fl.	,, 9	6	Mar. 12	May 5	May 1
	*House Sparrow (*Passer domesticus*), builds	,, 10	3	,, 22	Apr. 23	
	Burnet Rose (*Rosa spinosissima*), l.	,, 10	5	,, 16	,, 21	
10	*Wood Anemony (*Anemone nemorosa*), fl.	,, 10				
	Common Laurel (*Prunus laurocerasus*), fl.	,, 10	3	Mar. 20	Apr. 26	
	Wild Tulip (*Tulipa sylvestris*), fl.	,, 10				
	Goldfinch (*Carduelis elegans*), sg. com.	,, 10	7	Mar. 1	May 1	
11	Tit Pipit (*Anthus pratensis*), sg. com.	,, 11	5	,, 29	Apr. 28	
	Redbreast (*Erithaca rubecula*), lays	,, 11	3	Apr. 8	,, 14	
	*Common Lizard (*Zootoca vivipara*), ap.	,, 11	5	Mar. 12	,, 30	
	*Small-white Butterfly (*Pontia rapæ*), ap.	,, 11	7	,, 14	May 9	May 12
	*Humble-bee Fly (*Bombylius medius*), ap.	,, 11	7	Apr. 1	,, 1	
	*Alder (*Alnus glutinosa*), l.	,, 11	10	Mar. 19	Apr. 23	
	*Pear (*Pyrus communis*), fl.	,, 11	8	,, 19	,, 21	Apr. 24
	*Ash (*Fraxinus excelsior*), fl.	,, 11	11	,, 25	,, 27	,, 26

(a) This is in reference to the peculiar crow of the cock bird, when under the influence of sexual desire, and is only heard at the approach of, and during the breeding season : it is a good prognostic of warm spring weather.

(b) In 1834, this shrub was in flower on the 26th of January!

Apr.	Phenomena.	Date of occurrence; 1820—31.				Date in 1845.
		mean.		Earliest.	Latest.	
11	Heart's-ease (*Viola tricolor*), fl.	Apr. 11	6	Mar. 26	Apr. 19	May 5
	Field Wood-rush (*Luzula campestris*), fl.	,, 11	9	,, 26	May 1	Apr. 23
	*Red Currant (*Ribes rubrum*), fl.	,, 11	8	,, 31	Apr. 17	
12	Song Thrush (*Turdus musicus*), lays	,, 12	5	,, 21	May 12	
	*Lime (*Tilia europæa*), l.	,, 12	12	,, 27	Apr. 28	Apr. 23
	*Sycamore (*Acer pseudoplatanus*), l.	,, 12	12	,, 28	,, 25	,, 23
13	Stock-dove (*Columba œnas*), lays	,, 13	3	,, 27	,, 26	
	Carrion-Beetle (*Necrophorus humator*), ap. ...	,, 13	4	Apr. 2	,, 22	
	Redbreast (*Erithaca rubecula*), hatches.....	,, 14	3	Mar. 31	,, 22	
14	Wren (*Troglodytes europæus*), builds	,, 14	3	Apr. 4	,, 21	
	Chaffinch (*Fringilla cœlebs*), builds........	,, 14	3	,, 10	,, 16	
	*Wild Guelder-rose (*Viburnum opulus*), l. (a) ...	,, 14	5	,, 4	,, 22	
	*Cherry (*Prunus cerasus*), fl.	,, 14	7	Mar. 12	,, 29	
	Trailing Daphne (*Daphne cneorum*), fl.	,, 14				
15	*Willow Warbler (*Sylvia trochilus*), first heard (b)	,, 15	12	Apr. 5	Apr. 22	Apr. 22
	*Redstart (*Phœnicura ruticilla*), first heard....	,, 15	12	,, 6	,, 30	,, 23
	*Plume-footed Bee (*Anthophera retusa*), ap. (c) ...	,, 15	:	...	...	,, 23
	*Maple (*Acer campestre*), l.	,, 15	15	Mar. 25	May 1	,, 26
	*Hornbeam (*Carpinus betulus*), l.	,, 15	9	,, 28	Apr. 28	,, 22
	Strawberry-leaved Cinquefoil (*Potentilla fragariastrum*), fl.	,, 15	3	,, 21	,, 29	
16	*Blackcap (*Curruca atricapilla*), first heard	,, 16	12	,, 28	May 1	Apr. 23
	Long-eared Owl (*Otus vulgaris*), lays	,, 16				
	Silpha obscura, ap............................	,, 16				

			Apr. 16		
16 {	Oiceoptoma thoracica, ap.	Apr. 16	3	Apr. 14	Apr. 18	May 25
	*Common Snail (Helix aspersa), comes abroad	" 16	5	" 6	May 6	
	Black Slug (Arion ater), ap.	" 16				
	{	" 17				
17 {	†Dog Violet (Viola canina), fl.	" 16	7	Apr. 2	May 3	Apr. 2
	Great Plover (Œdicnemus crepitans), first heard or seen...	" 17	3	" 3	" 2	
	Frog Tadpoles, hatched	" 17	3	" 12	Apr. 22	
	*Black Currant (Ribes nigrum), fl.	" 17	2	" 14	" 20	Apr. 30
	Marsh Titmouse (Parus palustris), note ceases	" 18				
	Long-tailed Titmouse (P. caudatus), lays	" 18				
	Hedge Accentor (Accentor modularis), hatches	" 18				
18 {	Silpha lævigata, ap.	" 18	2	Apr. 16	Apr. 20	Jun. 3
	†Meadow Lady's-smock (Cardamine pratensis), fl.	" 18	11	Mar. 21	May 7	Apr. 26
	Wild Chervil (Anthriscus sylvestris), fl.	" 18	9	" 29	" 1	" 27
	Fritillary (Fritillaria meleagris), fl.	" 18	4	Apr. 9	Apr. 28	
	Peewit (Vanellus cristatus), lays [d]	" 19	3	Mar. 29	May 20	May 16
19 {	Blackbird (Turdus merula), hatches	" 19	4	Apr. 9	" 13	
	*Swallow (Hirundo rustica), first seen [e]	" 19	12	" 9	Apr. 26	Apr. 22

(a) In noting the periods of leafing and flowering in this shrub, it will be well in all cases to distinguish the wild sort from the large snow-ball variety cultivated in gardens.

(b) Though the 15th of April is given as the mean date of twelve years for this species of warbler being first heard, I do not find this date affected by taking the mean of twenty-one years. It may therefore be considered as the true mean date of this phenomenon. In like manner, the time given for the redstart's being first heard may be considered as the mean of sixteen years; that for the blackcap (Apr. 16) as the mean of nineteen; that for the tree-pipit (Apr. 20) as the mean of sixteen; that for the nightingale (Apr. 21) as the mean of twenty.

(c) This species of bee is not uncommon in gardens at Swaffham Bulbeck at this season, and is a good prognostic. It may often be observed hovering over the blossoms of the stinking hellebore (Helleborus fætidus), to which plant it seems much attached.

(d) It is this species which produces the "Plovers' eggs" eaten at table.

(e) The above mean date of Apr. 19, for the first arrival of the swallow, coincides with a mean of twenty years.

APR.	PHENOMENA.	Date of occurrence; 1820—31.				Date in 1845.
		Mean.		Earliest.	Latest.	
19	*Queen Wasp (*Vespa vulgaris*), ap.(ᵃ)	Apr. 19	8	Mar. 3	May 14	Apr. 20
	Sparkler-Beetle (*Cicindela campestris*), ap.	,, 19	5	,, 15	,, 6	Apr. 28
	*White Poplar (*Populus alba*), l.	,, 19	7	Apr 6	,, 2	
	Corn Horsetail (*Equisetum arvense*), fl.	,, 19	3	,, 12	,, 3	
	Song Thrush (*Turdus musicus*), hatches	,, 20	2	,, 6	,, 6	
20	Missel Thrush (*T. viscivorus*), hatches	,, 20	2	,, 10	Apr. 30	Apr. 29
	†Tree Pipit (*Anthus arboreus*), first heard	,, 20	12	,, 7	,, 29	Mar. 29
	Steropus madidus, ap.	,, 20	.	...	...	
	Pasque-flower (*Anemone pulsatilla*), fl.	,, 20	5	Apr. 1	May 6	Apr. 23
	*Birch (*Betula alba*), fl.	,, 20	6	,, 10	,, 1	,, 26
	*Nightingale (*Philomela luscinia*), first heard	,, 21	12	,, 8	Apr. 28	,, 24
21	Narrow-leaved Mouse-ear-chickweed (*Cerast. triviale*), fl.	,, 21	5	Feb. 28	May 5	
	Wood Crowfoot (*Ranunculus auricomus*), fl.	,, 21	8	Mar. 31	,, 5	
	*Hornbeam (*Carpinus betulus*), fl.	,, 21	5	Apr. 5	Apr. 30	Apr. 28
	Kestrel (*Falco tinnunculus*), lays	,, 22				
22	Ringed Snake (*Natrix torquata*), couples	,, 22				
	Pale Narcissus (*Narcissus biflorus*), fl.	,, 22	2	Apr. 19	Apr. 25	
	Jelly Nostoc (*Nostoc commune*), ap. on lawns	,, 22				
	Squirrel (*Sciurus vulgaris*), builds	,, 23				
	Young Hedge Accentors fledged	,, 23				
23	Young Moorhens hatched	,, 23				
	Swallow (*Hirundo rustica*), sg. com.	,, 23	8	Apr. 16	May 2	
	Jackdaw (*Corvus monedula*), lays	,, 23	3	,, 20	Apr. 27	

Phenomenon					
*Large-white Butterfly (*Pontia brassicæ*), ap.	Apr. 23	7	Mar. 26	May 11	Apr. 24
Black Poplar (*Populus nigra*), l.	,, 23	12	Apr. 10	May 3	Apr. 26
*Beech (*Fagus sylvatica*), l.	,, 23	7	,, 13	,, 2	Apr. 30
Lombardy Poplar (*Populus dilatata*), l.	,, 23	11	,, 13	,, 7	
†Jack-by-the-hedge (*Alliaria officinalis*), fl.	,, 23	4	,, 3	,, 8	
Morell (*Morchella esculenta*), ap.	,, 23	8	,, 18	Apr. 28	May 31
*Whinchat (*Saxicola rubetra*), first heard.	,, 24	7	,, 20	May 4	Apr. 22
*Wryneck (*Yunx torquilla*), first heard.	,, 24	4	,, 9	,, 8	26
Dogwood (*Cornus sanguinea*), l.	,, 24	7	,, 12	,, 5	
Water Crowfoot (*Ranunculus aquatilis*), fl.	,, 24	2	,, 19	Apr. 29	
*Butter-bur (*Petasites vulgaris*), fl.	,, 24	4	,, 2	May 10	
Linnet (*Linota cannabina*), lays.	,, 25	11	,, 14	,, 4	
*Whitethroat (*Curruca cinerea*), first heard.	,, 25	12	,, 15	,, 9	May 6
*Sedge Warbler (*Salicaria phragmitis*), first heard.	,, 25	4	,, 18	Apr. 30	,, 6
Ring-dove (*Columba palumbus*), lays.	,, 25	2	,, 16	May 5	
Tissue Moth (*Triphosa dubitata*), ap.	,, 25	2	,, 17	Apr. 29	Sept. 19
*Sialis lutarius, ap.	,, 25	4	,, 11	May 5	May 1
†Buttercup (*Ranunculus bulbosus*), fl.	,, 25	12	,, 11	,, 10	Apr. 29
Blue-bell (*Agraphis nutans*), fl.	,, 25	2	,, 19	,, 1	
German Iris (*Iris germanica*), fl.	,, 25	2			
Young Blackbirds fledged.	,, 26	.	...	...	
Nuthatch (*Sitta europæa*), whistling note heard.	,, 26				May 18
*Lesser Whitethroat (*Curruca sylviella*), first heard.	,, 26	12	Apr. 17	May 2	Apr. 24
Oiceoptoma rugosa, ap.	,, 26	2	,, 13	,, 10	
*Green-veined-white Butterfly (*Pontia napi*), ap.	,, 26	6	,, 15	,, 9	Apr. 19

(a) This entry is in reference to the large queen wasps which appear in spring, and which are the founders of new colonies. The workers which abound so in the summer and autumn are their offspring, and do not show themselves till much later.

Apr.	Phenomena.	Mean.		Earliest.	Latest.	Date in 1845.
26	Common Snail (*Helix aspersa*), engenders	Apr. 26				
	American-Cowslip (*Dodecatheon meadia*), fl.	,, 26				
	Henbit Dead-nettle (*Lamium amplexicaule*), fl.	,, 26	6	Apr. 6	May 22	Apr. 30
	*Strawberry (*Fragaria vesca*), fl.	,, 26	11	,, 10	,, 13	
	*Quince (*Cydonia vulgaris*), fl.	,, 26	4	,, 15	,, 18	
	Dotterel (*Charadrius morinellus*), makes its spring passage	,, 27	3	,, 19	,, 7	May 28
27	*Cuckoo (*Cuculus canorus*), first heard	,, 27	12	,, 21	,, 7	,, 1
	Poplar Hawkmoth (*Smerinthus populi*), ap.	,, 27	2	,, 13	,, 12	
	*Walnut (*Juglans regia*), l.	,, 27	12	,, 14	,, 9	Apr. 30
	Ribwort Plantain (*Plantago lanceolata*), fl.	,, 27	9	,, 15	,, 13	,, 29
	†Crab (*Pyrus malus*), fl.	,, 27	10	,, 19	,, 8	May 13
28	Reed Bunting (*Emberiza schœniclus*), sg. com	,, 28				
	Chaffinch (*Fringilla cœlebs*), lays	,, 28	6	Apr. 17	May 4	May 9
	*Beech (*Fagus sylvatica*), fl.	,, 28	4	,, 26	,, 3	
	Young Redbreasts fledged	,, 29	3	,, 22	,, 4	
	*Large-tortoiseshell Butterfly (*Vanessa polychloros*), ap.	,, 29	2			
29	Banded Snail (*Helix nemoralis*), engenders	,, 29	4	Apr. 17	May 12	
	Wich Elm (*Ulmus montana*), l.	,, 29	8	,, 23	,, 4	May 15
	*Barberry (*Berberis vulgaris*), fl.	,, 29	12	,, 19	,, 8	Apr. 29
	†Germander Speedwell (*Veronica chamædrys*), fl.	,, 29	9	,, 19	,, 13	May 11
	*Maple (*Acer campestre*), fl.	,, 29	8	,, 19	,, 13	,, 1
	*Sycamore (*A. pseudo platanus*), fl.	,, 29		,, 21	,, 4	
30	Ring-dove (*Columba palumbus*), hatches	,, 30				

*Martin (Hirundo urbica), first seen	Apr. 30	12	Apr. 15	May 19	May 24
Young Song Thrushes fledged	,, 30	2	,, 17	,, 13	
Speckled-wood Butterfly (Hipparchia egeria), ap.	,, 30	6	,, 20	,, 10	
Lime Hawkmoth (Smerinthus tiliæ), ap.	,, 30	3	,, 23	,, 10	
Dung of cattle swarms with coleopterous insects (a)	,, 30	3	,, 28	,, 2	
*Ash (Fraxinus excelsior), l.	,, 30	12	,, 22	,, 9	May 8
*Cuckow-pint (Arum maculatum), fl.	,, 30	9	,, 10	,, 12	,, 6
*Lilac (Syringa vulgaris), fl.	,, 30	10	,, 21	,, 13	,, 15
London-pride (Saxifraga umbrosa), fl.	,, 30	2	,, 29	,, 2	,, 29

(Rows above grouped under **30**.)

MAY.

Hister unicolor, ap.	May 1	2	Apr. 29	May 4	May 14
Fig (Ficus carica), l.	,, 1	5	,, 10	,, 18	,, 8
*Vine (Vitis vinifera), l.	,, 1	5	,, 24	,, 11	May 15
Young Rooks fledged	,, 2	11	,, 26	,, 9	
Rhingia rostrata, ap.	,, 2	2			
*Oak (Quercus robur), l.	,, 2	9	Apr. 25	May 12	May 17
Caper Spurge (Euphorbia lathyris), fl.	,, 2	5	,, 8	,, 25	
*Bladder-nut (Staphylea pinnata), fl.	,, 2	4	,, 23	,, 13	May 14
*Mealy-tree (Viburnum lantana), fl.	,, 3	9	,, 24	,, 13	
Yellow Wagtail (Motacilla raii), first seen	,, 3	6	,, 28	,, 7	
*Medlar (Mespilus germanica), fl.	,, 3	3			
Dove's-foot Cranesbill (Geranium molle), fl.	,, 3	9	April 12	May 16	
Black Medick (Medicago lupulina), fl.	,, 3	8	,, 15	June 1	Apr. 29
*Horse-chestnut (Æsculus hippocastanum), fl.	,, 3	11	,, 19	May 16	May 12

(Rows above grouped under **1**, **2**, **3**.)

a These consist principally of Aphodii (A. fossor, luridus, fimetarius, and others) and the smaller Staphylinidæ.

MAY	PHENOMENA.	Date of occurrence; 1820—31.			Date in 1845.
		Mean.		Earliest.	Latest.	

MAY	PHENOMENA.	Mean.		Earliest.	Latest.	Date in 1845.
3	*Bugle (*Ajuga reptans*), fl.	May 3	9	Apr. 2	May 15	May 13
	Chaffinch (*Fringilla cælebs*), hatches	,, 4	9	May 1	May 29	May 12
	*Pettychaps (*Curruca hortensis*), first heard ..	,, 4	3	,, 2	,, 6	
4	†Wood Warbler (*Sylvia sibilatrix*), first heard ..	,, 4	2	Apr. 27	,, 11	
	Mullein Moth (*Cucullia verbasci*), ap.	,, 4	10	Mar. 23	,, 18	May 24
	Herb-Robert (*Geranium robertianum*), fl.	,, 4	9	Apr. 15	,, 17	
	Field Chickweed (*Cerastium arvense*), fl.	,, 5				
	Latticed-heath Moth (*Hercyna clathrata*), ap. ..	,, 5				
	Empis pennipes ap.	,, 5	12	Apr. 19	May 20	May 13
5	*Whitethorn (*Cratægus oxyacantha*), fl.	,, 5	3	May 1	,, 12	,, 8
	*Woodruff (*Asperula odorata*), fl.	,, 5				
	*Common Elm sheds its seed	,, 6				
	Greenfinch (*Coccothraustes chloris*), builds	,, 6	3	Apr. 13	May 25	
	Creeper (*Certhia familiaris*), builds	,, 6	4	,, 23	,, 20	
6	Pheasant (*Phasianus colchicus*), lays	,, 6	9	,, 29	,, 18	May 22
	Plane (*Platanus orientalis*), l.	,, 6	11	,, 20	,, 19	,, 19
	Pale perfoliate Honeysuckle (*Lonicera caprifolium*), fl. ..	,, 7				
	Long-eared Bat (*Plecotus auritus*), comes abroad..	,, 7				
	Swallow (*Hirundo rustica*), builds........	,, 7				
7	*Sand-Martin (*H. riparia*), first seen	,, 7	6	Apr. 20	May 23	
	Creeper (*Certhia familiaris*), lays	,, 7	2	,, 26	,, 18	
	Mountain-ash (*Pyrus aucuparia*) fl.	,, 7				
	*Red Clover (*Trifolium pratense*) fl.	,, 7	10	Apr. 25	May 18	May 13

	Phenomenon					
	Common Fumitory (*Fumaria officinalis*), fl.	May 7	9	Apr. 25	Jun. 12	May 31
7	Bitter Winter-cress (*Barbarea vulgaris*) fl.	,, 7	7	,, 27	May 19	,, 14
	*Persian Lilac (*Syringa persica*), fl.	,, 7	4	,, 28	,, 18	
	*Turtle-dove (*Columba turtur*), first heard	,, 8	9	,, 27	,, 15	May 6
	Greenfinch (*Coccothraustes chloris*), lays	,, 8				
	Emperor Moth (*Saturnia pavonia-minor*), ap.	,, 9	3	Apr. 17	Jun. 4	Jun. 2
	*Orange-tip Butterfly (*Pontia cardamines*), ap.	,, 8	11	May 2	May 18	
	Small marsh Valerian (*Valeriana dioica*), fl.	,, 8	7	Apr. 20	,, 28	
8	*Oak (*Quercus robur*), fl.	,, 8	7	,, 30	,, 22	Jun. 2
	Great Carex (*Carex riparia*), fl.	,, 8	3	May 1	,, 12	
	Burying-Beetle (*Necrophorus vespillo*), ap.	,, 9	3	Apr. 24	Jun. 2	
	Flesh-fly (*Sarcophaga carnaria*), ap.	,, 9		May 3	May 14	
	Thyme-leaved Speedwell (*Veronica serpyllifolia*), fl.	,, 9	7	,, 3	,, 15	May 15
	White Jasmine (*Jasminum officinale*), l.	,, 9	2	,, 5	,, 17	,, 13
9	Yellow Fumitory (*Corydalis lutea*), fl.	,, 10	3			
	Noctule Bat (*Vespertilio noctula*), comes abroad	,, 10	2	May 6	May 14	
	Partridge (*Perdix cinerea*), lays	,, 10		May 7		
	Pale tussock Moth (*Dasychira pudibunda*), ap.	,, 10	2	,, 8	May 13	
	Buff-tip Moth (*Pygæra bucephala*), ap.	,, 10	2	,, 2	,, 12	May 27
10	*Harry-long-legs (*Tipula oleracea*), ap.	,, 10	6	,, 8	,, 23	,, 31
	*Walnut (*Juglans regia*), fl.	,, 10	4	May 3	,, 12	
	Shepherd's-needle (*Scandix pecten-veneris*), fl.	,, 11	3	Apr. 16	May 20	May 15
	Reed Bunting (*Emberiza schœniclus*), lays	,, 11	3	,, 30	Jun. 20	Jun. 3
11	*Aspen (*Populus tremula*), l.	,, 11		May 3	May 20	
	Hemlock Stork's-bill (*Erodium cicutarium*), fl.	,, 11	3	Apr. 16	Jun. 20	May 15
	Lesser Burnet (*Poterium sanguisorba*), fl.	,, 11	5	,, 30	May 20	Jun. 3

MAY	PHENOMENA.	Date of occurrence; 1820—31. Mean.		Date of occurrence; 1820—31. Earliest.	Date of occurrence; 1820—31. Latest.	Date in 1845.
11	*Celandine (*Chelidonium majus*), fl.	May 11	6	May 1	May 25	May 24
	†Water-violet (*Hottonia palustris*), fl.	,, 11	7	,, 1	,, 30	May 23
	Charlock (*Sinapis arvensis*), fl.	,, 11	6	,, 2	,, 20	,, 24
	*Laburnum (*Cytisus laburnum*), fl.	,, 11	7	,, 5	,, 17	
	Long-tailed Titmouse (*Parus caudatus*), hatches	,, 12				
	†Wall Butterfly (*Hipparchia megæra*), ap.	,, 12	7	May 4	May 19	
	Black Bog-rush (*Schœnus nigricans*), fl.	,, 12				
12	Milk-wort (*Polygula vulgaris*), fl.	,, 12	6	Apr. 20	May 28	May 31
	Corn Gromwell (*Lithospermum arvense*), fl.	,, 12	4	May 5	,, 17	
	* Comfrey (*Symphytum officinale*), fl.	,, 12	6	,, 5	,, 22	
	Green-winged meadow Orchis (*Orchis morio*), fl.	,, 12	3	,, 9	,, 19	
	*Lily-of-the-valley (*Convallaria majalis*), fl.	,, 12	5	,, 9	,, 20	May 29
	Willow Warbler (*Sylvia trochilus*), lays.	,, 13				
13	*Swift (*Cypselus apus*), first seen.	,, 13	12	May 6	May 30	
	Nightingale (*Philomela luscinia*), lays.	,, 13	2	,, 8	,, 18	
	*Scorpion-fly (*Panorpa communis*), ap.	,, 13	4	,, 11	,, 25	
	†English Clary (*Salvia verbenacea*), fl.	,, 13	11	,, 1	Jun. 1	May 15
	Corn-sallad (*Valerianella olitoria*), fl.	,, 13	2	,, 5	May 21	
	Snowdrop Anemony (*Anemone sylvestris*), fl.	,, 13	2	,, 7	,, 20	
	Columbine (*Aquilegia vulgaris*), fl.	,, 13	5	,, 8	,, 17	
	Meadow Fox-tail-grass (*Alopecurus pratensis*), fl.	,, 13	6	,, 8	,, 18	May 28
	Yellow Archangel (*Lamium galeobdolon*), fl.	,, 13	2	,, 8	,, 18	Jun. 2
14	Tit Pipit (*Anthus pratensis*), lays.	,, 14				

Phenomenon	Date	No.			
Dingy-skipper Butterfly (*Thymele tages*), ap.	May 14		May 10	May 18	
Dot Moth (*Mamestra persicariae*), ap.	„ 14	2	Apr. 12	„ 28	Jun. 2
Blue Sherardia (*Sherardia arvensis*), fl.	„ 14	5	„ 30	Jun. 3	
Common Hedge-mustard (*Sisymbrium officinale*), fl.	„ 14	6	May 9	May 19	
Water Avens (*Geum rivale*), fl.	„ 14	2	„ 11	„ 21	
Fly Orchis (*Ophrys muscifera*), fl.	„ 14	4	„ 12	„ 17	
Sweet-scented Vernal-grass (*Anthoxanthum odoratum*), fl.	„ 14	2	„ 4	„ 22	
Yellow-hammer (*Emberiza citrinella*), lays.	„ 15	3	Apr. 14	„ 4	Jun. 13
*Cockchaffer (*Melolontha vulgaris*), ap.	„ 15	7	May 9	Jun. 7	
Soldier-Beetle (*Telephorus lividus*), ap.	„ 15	2		May 20	„ 3
Star-of-Bethlehem (*Ornithogalum umbellatum*), fl.	„ 15		May 6	May 24	Jun. 2
Paeony (*Paeonia officinalis*), fl.	„ 15	4	„ 7	„ 21	
Solomon's-seal (*Convallaria multiflora*), fl.	„ 15	3			
Lesser Whitethroat (*Curruca sylviella*), lays.	„ 16	8	May 12	May 24	
*Spotted Flycatcher (*Muscicapa grisola*), first seen.	„ 16	2	„ 4	„ 29	
Harpalus ruficornis, ap.	„ 16	5	„ 1	„ 28	
Common Tormentil (*Potentilla tormentilla*), fl.	„ 16	4	„ 1	„ 28	
Celery-leaved Crowfoot (*Ranunculus sceleratus*), fl.	„ 16	7	„ 4	Jun. 4	
*Upright Crowfoot (*R. acris*), fl.	„ 16			May 30	May 29
Young broods of Chaffinches fledged.	„ 17	7	Apr. 29	Jun. 9	
*May-fly (*Ephemera vulgata*), ap.	„ 17				Jun. 2
*Alder Buckthorn (*Rhamnus frangula*), fl.	„ 17	5	May 14	May 21	
Tway-blade (*Listera ovata*), fl.	„ 17				
Tree Pipit (*Anthus arboreus*), lays.	„ 18				
Goldfinch (*Carduelis elegans*), lays.	„ 18	7	Mar. 27	July 12	
Midge (*Thrips physapus*), ap.	„ 18				
Buck-bean (*Menyanthes trifoliata*), fl.	„ 18				

MAY	PHENOMENA.	Date of occurrence; 1820—31.			Date in 1845.
		Mean.	Earliest.	Latest.	
18	Field Flea-wort (*Cineraria campestris*), fl........	May 18 ³	May 3	Jun. 7	
	Field Scorpion-grass (*Myosotis arvensis*), fl.	,, 18 ⁶	,, 7	,, 2	
	*Monk's-hood (*Aconitum napellus*), fl........	,, 18 ³	,, 8	May 24	Jun. 27 4
	†Silver-weed (*Potentilla anserina*), fl.	,, 18 ⁷	,, 13	,, 31	,,
19	Young broods of Starlings fledged.	,, 19			
	Blackcap (*Curruca atricapilla*), lays.	,, 19			
	Young broods of Greenfinches hatched	,, 19 ³	May 13	May 28	May 27
	*Mulberry (*Morus nigra*), l........	,, 19 ⁴	,, 12	,, 28	
	†Holly (*Ilex aquifolium*), fl.	,, 19 ⁴	,, 5	Jun. 10	
	†Ragged-robin (*Lychnis flos-cuculi*), fl........	,, 19 ⁶	,, 10	,, 9	
20	Sedge Warbler (*Salicaria phragmitis*), lays.	,, 20			
	Whinchat (*Saxicola rubetra*), lays.	,, 20			Jun. 2
	*Sailor-Beetle (*Telephorus rusticus*), ap.	,, 20	...	...	
21	Cross-wort (*Galium cruciatum*), fl.	,, 21 ⁴	May 12	May 26	May 31
	†House-martin (*Hirundo urbica*), builds........	,, 21			
	*Golden-green Dragon-fly (*Calepteryx virgo*), ap.	,, 21			
	Malachius aeneus, ap........	,, 21			Jun. 2
22	Soft Brome-grass (*Serrafalcus mollis*), fl........	,, 22 ⁸	May 5	Jun. 14	
	Great Titmouse (*Parus major*), sg. ceas........	,, 22 ⁶	,, 3	,, 8	Jun. 3
	*Raspberry (*Rubus idaeus*), fl........	,, 22 ⁷	,, 3	,, 9	May 31
	White Clover (*Trifolium repens*), fl........	,, 22 ⁵	,, 4	,, 8	Jun. 3
	Wild Mignonette (*Reseda lutea*), fl.	,, 22 ⁶	,, 6	,, 3	,, 4
	White Campion (*Lychnis vespertina*), fl.	,, 22			

Species	Date	No.			
Greasy-Fritillary Butterfly (*Melitæa artemis*), ap.	May 23	…	…	…	Jun. 2
Sea Pancratium (*Pancratium maritimum*), fl.	„ 23	5	May 3	Jun. 7	„ 3
Purple mountain Milk-vetch (*Astragalus hypoglottis*), fl.	„ 23	8	„ 14	„ 4	„ 6
†Herb-Bennet (*Geum urbanum*), fl.	„ 23	5	„ 16	May 29	
Common Gromwell (*Lithospermum officinale*), fl.	„ 23	9	„ 14	Jun. 11	
†Small-heath Butterfly (*Hipparchia pamphilus*), ap.	„ 24	2	„ 19	May 29	
Heath Moth (*Fidonia atomaria*), ap.	„ 24	6	Apr. 30	Jun. 22	Jun. 3
Slender Fox-tail-grass (*Alopecurus agrestis*), fl.	„ 24	5	May 1	May 31	Jun. 3
†Bird's-foot Trefoil (*Lotus corniculatus*), fl.	„ 24	5	„ 2	Jun. 6	Jun. 3
Wild Radish (*Raphanus raphanistrum*), fl.	„ 24	6	„ 3	„ 3	May 30
Dwarf dark-winged Orchis (*Orchis ustulata*), fl.	„ 24	6	„ 16	„ 7	
Mouse-ear Hawkweed (*Hieracium pilosella*), fl.	„ 24	6	„ 14	„ 12	
†Hive Bee swarms	„ 25	…	…	…	Jun. 4
Butterwort (*Pinguicula vulgaris*), fl.	„ 25	…	…	Jun. 12	
Hairy Thrincia (*Thrincia hirta*), fl.	„ 25	2	May 7	„ 22	
*Jacob's-ladder (*Polemonium cæruleum*), fl.	„ 25	4	„ 8	„ 4	Jun. 13
*Bistort (*Polygonum bistorta*), fl.	„ 25	5	„ 16	„ 13	
Dewberry (*Rubus cæsius*), fl.	„ 25	4	„ 18	„ 3	
Grizzled-Skipper Butterfly (*Thymele alveolus*), ap.	„ 26	2	„ 19	„ 4	Jun. 3
Brown-argus Butterfly (*Polyommatus agrestis*), ap.	„ 26	5	„ 4	„ 20	
Tufted Horse-shoe-vetch (*Hippocrepis comosa*), fl.	„ 26	3	„ 9	„ 9	
Early purple Orchis (*Orchis mascula*), fl.	„ 26	9	„ 9	„ 12	Jun. 4
*Guelder-rose (*Viburnum opulus*), fl.	„ 26	6	„ 12	„ 16	„ 3
Common Rock-rose (*Helianthemum vulgare*), fl.	„ 26	7	„ 16	„ 7	
Missel Thrush (*Turdus viscivorus*), sg. ceas.	„ 27	.	„ 3	„ 24	
Garden Carpet-Moth (*Cidaria fluctuata*), ap.	…	.	…	…	
Wood Scorpion-grass (*Myosotis sylvatica*), fl.	„ 27	.	…	…	May 27

MAY	PHENOMENA.	Mean.		Date of occurrence; 1820—31. Earliest.	Latest.	Date in 1845.
	Evergreen Oak (Quercus ilex), fl.	May 27	2	May 7	Jun. 16	
27	Brooklime (Veronica beccabunga), fl.	,, 27	7	,, 14	,, 11	Jun. 4
	*Ox-eye-Daisy (Chrysanthemum leucanthemum), fl.	,, 27	9	,, 17	,, 11	
	Hoary Plantain (Plantago media), fl.	,, 27	6	,, 18	,, 9	Jun. 6
	Young broods of Linnets fledged		:	...	...	May 28
28	Common Sandpiper (Totanus hypoleucos), first seen	May 28		...	...	
	Smooth-stalked Meadow-grass (Poa pratensis), fl.	,, 28	6	May 13	Jun. 20	
	Wood Sanicle (Sanicula europæa), fl.	,, 28				
	Hound's-tongue (Cynoglossum officinale), fl.	,, 28	7	,, 18	,, 14	May 20
	Common Sorrel (Rumex acetosa), fl.	,, 28	5	,, 1	July 16	
	*Quail (Coturnix vulgaris), note first heard	,, 29	8	,, 9	,, 1	Jun. 6
29	Goose-grass (Galium aparine), fl.	,, 29	3	,, 25	Jun. 8	,, 9
	*Yellow Day-lily (Hemerocallis flava), fl.	,, 29	3	,, 9	,, 15	,, 3
	*Stinging-fly (Stomoxys calcitrans), ap.	,, 29	9	,, 11	,, 28	
	Swallow-tail Butterfly (Papilio machaon), ap.	,, 30	5	,, 13	,, 15	
30	*Libellula depressa, ap.	,, 30	4	,, 19	,, 5	
	Corn Crowfoot (Ranunculus arvensis), fl.	,, 30	5	,, 20	,, 12	
	Yellow Rattle (Rhinanthus crista-galli), fl.	,, 30	4	,, 21	,, 7	
	Marsh Lousewort (Pedicularis palustris), fl.	,, 30	2	,, 29	,, 3	
	Young broods of Whitethroats (Curruca cinerea), fledged	,, 31				
31	Four-spotted Dragon-fly (Libellula quadrimaculata), ap.	,, 31	6	,, 17	July 4	Jun. 13
	Puss Moth (Cerura vinula), ap.	,, 31	2	,, 21	Jun. 10	
	Flat-stalked Meadow-grass (Poa compressa), fl.	,, 31				

31	Small Nettle (*Urtica urens*), fl.	May 31	5	Apr. 9	July 19	Jun. 9
	*Marsh Orchis (*Orchis latifolia*), fl.	,, 31	5	May 5	Jun. 28	,, 13
	*Red Bryony (*Bryonia dioica*), fl.	,, 31	7	,, 20	,, 9	,, 13
	*Syringa (*Philadelphus coronarius*), fl.	,, 31	8	,, 22	,, 10	,, 16

JUNE.

1	*Common Elder (*Sambucus nigra*), fl.	Jun. 1	12	May 20	Jun. 11	Jun. 13
2	*Virginian Spider-wort (*Tradescantia virginica*), fl.	,, 2	:	...	...	,, 13
	*Fraxinella (*Dictamnus alba*), fl.	,, 3	3	...	...	,, 16
3	Thyme-leaved Sandwort (*Arenaria serpyllifolia*), fl.	,, 3	6	May 13	July 10	Jun. 11
	Common red Poppy (*Papaver rhœas*), fl.	,, 3	7	,, 22	Jun. 15	
	Spotted Flycatcher (*Muscicapa grisola*), lays	,, 4	4	,, 24	,, 14	
	Redbreast lays a second time.	,, 4	2	,, 29	,, 9	
	Pyrochroa rubens, ap.	,, 4	7	,, 17	July 14	
	Clustered Bell-flower (*Campanula glomerata*), fl.	,, 4	6	,, 3	Jun. 26	
	Cow-parsnep (*Heracleum sphondylium*), fl.	,, 4	5	,, 7	July 20	
4	Water-cress (*Nasturtium officinale*), fl.	,, 4	6	,, 12	,, 15	July 15
	Rye-grass (*Lolium perenne*), fl.	,, 4	6	,, 20	Jun. 15	,, 8
	Rough Chervil (*Chærophyllum temulum*), fl.	,, 4	6	,, 21	,, 13	
	Downy Oat-grass (*Avena pubescens*), fl.	,, 4	2	,, 28	,, 12	Jun. 4
	Burnet Rose (*Rosa spinosissima*), fl.	,, 4	2	,, 31	,, 8	
	Creophilus maxillosus, ap.	,, 5	4	Apr. 17	July 23	Jun. 12
	*Common-blue Butterfly (*Polyommatus alexis*), ap.	,, 5	8	May 19	Jun. 28	
5	Small garden Chaffer (*Anomala horticola*), ap.	,, 5	2	,, 30	,, 11	
	Italian Bugloss (*Anchusa italica*), fl.	,, 5	.	...	...	
	*Deadly Nightshade (*Atropa belladonna*), fl.	,, 5	5	Jun. 1	Jun. 13	Jun. 12

JUNE	PHENOMENA.	Mean.		Earliest.	Latest.	Date in 1845.
5	*Rye (*Secale cereale*), fl.	Jun. 5	2	Jun. 3	Jun. 7	
	*Landrail (*Crex pratensis*), note first heard	,, 6	3	May 13	July 16	
	Young broods of Pheasants hatched	,, 6	3	,, 29	Jun. 18	
6	Spotted palmate Orchis (*Orchis maculata*), fl.	,, 6	7	May 18	Jun. 25	Jun. 17
	Bladder-campion (*Silene inflata*), fl.	,, 6	4	,, 23	,, 29	
	Long-prickly-headed Poppy (*Papaver argemone*), fl.	,, 6	6	,, 28	,, 16	
	Great Nettle (*Urtica dioica*), fl.	,, 7	9	,, 24	,, 15	Jun. 13
	Nightingale (*Philomela luscinia*), sg. ceas.	,, 7	.	....	....	,, 12
7	*Large Oriental Poppy (*Papaver orientale*), fl.	,, 7	6	Apr. 25	July 18	
	Flixweed (*Sisymbrium sophia*), fl.	,, 7	4	May 14	Jun. 29	
	Mountain Cudweed (*Antennaria dioica*), fl.	,, 7	6	,, 19	,, 26	
	Purging Flax (*Linum catharticum*), fl.	,, 7	5	,, 21	,, 22	
	Cock's-foot-grass (*Dactylis glomerata*), fl.	,, 7	6	,, 23	,, 26	
	*Saintfoin (*Onobrychis sativa*), fl.	,, 7	6	Jun. 5	,, 9	Jun. 17
	*Buckthorn (*Rhamnus catharticus*), fl.	,, 7	2	....	....	
	Red Valerian (*Centranthus ruber*), fl.	,, 8	6	May 7	Jun. 30	Jun. 9
8	Scarlet Pimpernel (*Anagallis arvensis*), fl.	,, 8	5	,, 12	,, 20	
	*Common Honeysuckle (*Lonicera periclymenum*), fl.	,, 8	4	,, 23	,, 24	Jun. 26
	Common Vetch (*Vicia sativa*), fl.	,, 8	4	Jun. 1	,, 20	
	Jagged-leaved Crane's-bill (*Geranium dissectum*), fl.	,, 8	2	,, 6	,, 10	
	Common Speedwell (*Veronica officinalis*), fl.	,, 9	5	May 15	July 9	
9	Dagger Moth (*Acronycta psi*), ap.	,, 9	5	,, 22	Jun. 30	July 20
	Dogwood (*Cornus sanguinea*), fl.	,, 9	6	,, 22	,, 30	Jun. 25

	Phenomenon					
9	*Small Bindweed (*Convolvulus arvensis*), fl.	Jun. 9	6	May 25	Jun. 28	Jun. 16
	†Dog-rose (*Rosa canina*), fl.	,, 9	9	,, 31	,, 22	,, 16
	Spotted Flycatcher (*Muscicapa grisola*), hatches	,, 10				
	Bright-line brown-eye Moth (*Mamestra oleracea*), ap.	,, 10				
	*Silver Y Moth (*Plusia gamma*), ap.	,, 10	5	May 5	July 18	
	Large brown Dragon-fly (*Æshna grandis*), ap.	,, 10	3	,, 17	,, 10	
	Hard Rush (*Juncus glaucus*), fl.	,, 10				
10	Butterfly Orchis (*Habenaria bifolia*), fl.	,, 10	7	May 7	July 12	July 7
	Wild Parsnep (*Pastinaca sativa*), fl.	,, 10	4	,, 30	Jun. 22	
	*Mulberry (*Morus nigra*), fl.	,, 10	6	,, 30	July 3	
	Yellow Flag (*Iris pseudacorus*), fl.	,, 10	4	Jun. 8	Jun. 12	
	Black Bryony (*Tamus communis*), fl.	,, 10				
	Young Jackdaws fledged	,, 11				
11	Dwarf Mallow (*Malva rotundifolia*), fl.	,, 11	5	May 16	Jun. 30	Jun. 25
	*Common Mallow (*M. sylvestris*), fl.	,, 11	6	Jun. 7	,, 18	,, 11
	Bloody Crane's-bill (*Geranium sanguineum*), fl.	,, 11	4	May 19	,, 27	Jun. 13
	White Mustard (*Sinapis alba*), fl.	,, 11	4	,, 23	,, 25	July 9
	*Wild Thyme (*Thymus serpyllum*), fl.	,, 11	6	,, 23	,, 27	Jun. 26
	Corn Blue-bottle (*Centaurea cyanus*), fl.	,, 11	6	,, 23	,, 29	
	*Great Snapdragon (*Antirrhinum magnus*), fl.	,, 11	5	,, 25	July 1	
	Lesser Spearwort (*Ranunculus flammula*), fl.	,, 11	4	,, 27	Jun. 20	
	Yellow Vetchling (*Lathyrus aphaca*), fl.	,, 11	2	Jun. 1	,, 21	
	†Hedge Woundwort (*Stachys sylvatica*), fl.	,, 11	6	,, 1	,, 29	
	Small Bugloss (*Lycopsis arvensis*), fl.	,, 11	3	,, 8	July 5	
	Meadow Hay cut	,, 11			Jun. 25	
12	Second broods of Redbreasts hatched	,, 12				
	Small elephant Hawkmoth (*Deilephila porcellus*), ap.	,, 12	12	May 30		Jun. 24

June	Phenomena	Mean		Earliest	Latest	Date in 1845
12	Curled Dock (*Rumex crispus*), fl.	Jun. 12	5	May 20	July 10	July 4
	Musk Thistle (*Carduus nutans*), fl. (a)	„ 12	6	„ 24	„ 16	
	Rough-stalked Meadow-grass (*Poa trivialis*), fl.	„ 12	4	Jun. 2	Jun. 23	
	Gout-weed (*Ægopodium podagraria*), fl.	„ 12	2	„ 9	„ 15	
13	Redstart (*Phœnicura ruticilla*), sg. ceas.	„ 13	3	May 27	July 11	Jun. 27
	Small-blue Butterfly (*Polyommatus alsus*), ap.	„ 13	6	Jun. 3	Jun. 19	
	Great-hedge Bedstraw (*Galium mollago*), fl.	„ 13	6	May 23	„ 30	
	Hazel-leaved Bramble (*Rubus corylifolius*), fl.	„ 13	5	Jun. 3	„ 15	
	*Woody Nightshade (*Solanum dulcamara*), fl.	„ 13	9	„ 3	„ 27	
	Upright Brome-grass (*Bromus erectus*), fl.	„ 13	3	„ 11	„ 15	
	Young broods of Swallows fledged	„ 14				
	Landrail (*Crex pratensis*), lays	„ 14				
14	Pink-underwing Moth (*Callimorpha jacobeæ*), ap.	„ 14	4	May 19	July 4	
	Large-skipper Butterfly (*Pamphila sylvanus*), ap.	„ 14	5	Jun. 6	Jun. 29	
	Moss Rose (*Rosa muscosa*), fl.	„ 14				
	*Frog-bit (*Hydrocharis morsus-ranæ*), fl.	„ 14	6	May 16	July 30	
	Lady's-fingers (*Anthyllis vulneraria*), fl.	„ 14	6	„ 22	„ 14	
	Meadow-rue (*Thalictrum flavum*), fl.	„ 14	6	„ 26	„ 16	
	Rough Hawk-bit (*Leontodon hispidum*), fl.	„ 14	3	„ 26	„ 5	
	Floating Meadow-grass (*Glyceria fluitans*), fl.	„ 14		Jun. 2	Jun. 16	
	Long-rooted Cat's-ear (*Hypochæris maculata*), fl.	„ 14	2	„ 12		
15	Young broods of Redstarts fledged	„ 15	2	„ 6	„ 25	
	Mare's-tail (*Hippuris vulgaris*), fl.	„ 15				

Worm-seed Treacle-mustard (*Erysimum cheiranthoides*), fl..	Jun. 15	7	Apr. 14	July 27	July 2
†Black Knapweed (*Centaurea nigra*), fl.	,, 15	6	May 22	Jun. 29	
Henbane (*Hyoscyamus niger*), fl.	,, 15	5	,, 22	,, 30	
Quaking-grass (*Briza media*), fl.	,, 15	4	Jun. 2	,, 23	
Melilot (*Melilotus officinalis*), fl.	,, 15	4	,, 4	,, 26	
Oat-grass (*Arrhenatherum avenaceum*), fl.	,, 15	2	,, 13	,, 17	
Wild Chamomile (*Matricaria chamomilla*), fl.	,, 16				
Corn Bell-flower (*Specularia hybrida*), fl.	,, 16	4	May 24	Jun. 30	Jun. 9
Sow-thistle (*Sonchus oleraceus*), fl.	,, 16	6	,, 24	July 2	,, 13
Trailing Dog-rose (*Rosa arvensis*), fl.	,, 16	5	Jun. 8	Jun. 22	
Common Wart-cress (*Senebiera coronopus*), fl.	,, 16	4	,, 9	,, 21	
Creeping Cinque-foil (*Potentilla reptans*), fl.	,, 16	5	,, 10	,, 20	
Narrow-leaved Oat-grass (*Avena pratensis*), fl.	,, 16	2	,, 12	,, 20	
Ivy casts its leaves	,, 16	3	,, 9	,, 29	
*Milk-thistle (*Silybum marianum*), fl.	,, 17				
*Good-King-Henry (*Chenopodium bonus-henricus*), fl.	,, 17	4	May 31	Aug. 5	
*Meadow Crane's-bill (*Geranium pratense*), fl.	,, 17	5	Jun. 1	July 4	
Dyer's Rocket (*Reseda luteola*), fl.	,, 17	5	,, 8	,, 10	
Sweet-William (*Dianthus barbatus*), fl.	,, 17	4	,, 14	Jun. 20	
Turtle-dove (*Columba turtur*), lays	,, 18	2			
Frog tadpoles nearly full grown, and acquiring fore feet..	,, 18	.	Jun. 17	Jun. 20	Jun. 27
Turk's-cap Lily (*Lilium martagon*), fl.	,, 18	5	...	...	
Corn-cockle (*Lychnis githago*), fl.	,, 18	5	May 23	July 7	
Forget-me-not (*Myosotis palustris*), fl.	,, 18		Jun. 7	,, 4	

Group brackets: 15, 16, 17, 18.

(*) Linnæus states in respect of the Thistles in general, " *Cardui varii non florent antequam solstitium absolutum est.*" (Philos. Bot.) This species, however, is an exception, and appears to be very uncertain in its time of flowering : in the present year (1846) two or three plants were found coming into flower so early as the 11th of March.

June	Phenomena.	Mean.		Earliest.	Latest.	Date in 1845.
				Date of occurrence; 1820—91		
18	Creeping Spike-rush (*Eleocharis palustris*), fl.	Jun. 18	3	Jun. 11	Jun. 30	
19	*Meadow-brown Butterfly (*Hipparchia janira*), ap.	,, 19	5	,, 9	,, 30	July 9
	Atopa cervina, ap.	,, 19	3	,, 14	,, 26	
	Pellitory-of-the-wall (*Parietaria officinalis*), fl.	,, 19	6	May 9	July 20	
	†Water Betony (*Scrophularia aquatica*), fl.	,, 19	4	,, 24	,, 4	Jun. 26
	Water Speedwell (*Veronica anagallis*), fl.	,, 19	5	Jun. 7	,, 9	
	Bee Orchis (*Ophrys apifera*), fl.	,, 19	7	,, 9	Jun. 28	
	Marsh Thistle (*Carduus palustris*), fl.	,, 19	3	,, 14	,, 28	Jun. 28
20	Young broods of Greenfinches fledged	,, 20				
	Eyed Hawk-moth (*Smerinthus ocellata*), ap.	,, 20				
	Small Scabious (*Scabiosa columbaria*), fl.	,, 20	3	Jun. 7	July 5	
	Sweet Milk-vetch (*Astragalus glycyphyllos*), fl.	,, 20				Jun. 30
	Self-heal (*Prunella vulgaris*), fl.	,, 20	5	Jun. 9	Jun. 28	
	Viper's Bugloss (*Echium vulgare*), fl.	,, 20	5	,, 10	,, 30	
	Wild Carrot (*Daucus carota*), fl.	,, 20	4	,, 14	July 2	
21	Young broods of Partridges hatched	,, 21				
	Asparagus-Beetle (*Crioceris asparagi*), ap.	,, 21	2	Jun. 2	July 10	
	Rose Beetle (*Cetonia aurata*), ap.	,, 21	2	,, 17	Jun. 26	
	Common Cudweed (*Filago germanica*), fl.	,, 21	5	,, 3	July 15	
	Goat's-beard (*Tragopogon pratensis*), fl.	,, 21	4	,, 7	,, 10	
	*Fox-glove (*Digitalis purpurea*), fl.	,, 21	2	,, 10	,, 2	
	Sweet-briar (*Rosa rubiginosa*), fl.	,, 21	2	,, 15	Jun. 27	
	Sweet-smelling Orchis (*Gymnadenia conopsea*), fl.	,, 21	3	,, 18	,, 24	

No.	Species					
21	Smooth Hawk's-beard (Crepis virens), fl.	Jun. 21				
22	Six-spot-Burnet Moth (Anthrocera filipendulæ), ap.	„ 22	5	Apr. 20	July 19	Jun. 27
	Broad-leaved Dock (Rumex obtusifolius), fl.	„ 22	4	Jun. 5	„ 4	„ 25
	*Biting Stonecrop (Sedum acre), fl.	„ 22	5	„ 9	„ 12	„ 27
	Nipplewort (Lapsana communis), fl.	„ 22	4	„ 13	Jun. 30	„ 25
	*Strawberries ripe.	„ 22	6	„ 2	July 5	
23	Gold-crested Wren (Regulus cristatus), sg. ceas.	„ 23	5	May 27	„ 16	
	Redbreast (Erithaca rubecula), sg. ceas.	„ 23	4	Jun. 9	„ 11	Jun. 28
	Enchanter's-Nightshade (Circæa lutetiana), fl.	„ 23	5	„ 10	„ 4	„ 26
	*Wheat (Triticum hybernum), fl.	„ 23	10	„ 10	„ 11	
	Meadow Vetchling (Lathyrus pratensis), fl.	„ 23	5	„ 12	„ 8	
	†Stinking Horehound (Ballota nigra), fl.	„ 23	7	„ 12	„ 11	
	Crested Dog's-tail-grass (Cynosurus cristatus), fl.	„ 23	2	„ 17	Jun. 29	
	Painted-lady Butterfly (Vanessa cardui), ap.	„ 24	2	„ 20	„ 28	
	Wild Oat (Avena fatua), fl.	„ 24				
24	Barren Brome-grass (Bromus sterilis), fl.	„ 24	2	Jun. 8	July 10	
	Hop-Trefoil (Trifolium procumbens), fl.	„ 24	5	„ 11	„ 12	
	May-weed (Anthemis cotula), fl.	„ 24				
	Portugal Laurel (Prunus lusitanica), fl.	„ 24				
25	*Common Wasp (Vespa vulgaris), begins to abound	„ 25	3	Jun. 18	Jun. 29	July 6
	*Small Horse-fly (Hæmatopota pluvialis), ap.	„ 25				
	Crested Hair-grass (Koeleria cristata), fl.	„ 25				
	*Tawny Day-lily (Hemerocallis fulva), fl.	„ 25	5	Jun. 17	July 9	
	*Dropwort (Spiræa filipendula), fl.	„ 25	3	„ 22	Jun. 30	
	Bee Larkspur (Delphinium elatior), fl.	„ 25				

JUNE	PHENOMENA.	Date of occurrence; 1820—31.				Date in 1845.
		Mean.		Earliest.	Latest.	
26	Privet Hawkmoth (*Sphinx ligustri*), ap.	Jun. 26	5	Jun. 16	July 4	July 8
	Strawberry Trefoil (*Trifolium fragiferum*), fl.	„ 26	5	May 28	„ 20	
	*Yellow Water-lily (*Nuphar lutea*), fl.	„ 26	5	„ 30	„ 26	
	Bull-rush (*Scirpus lacustris*), fl.	„ 26	2	Jun. 13	„ 9	
	Spotted Cat's-ear (*Achyrophorus maculatus*), fl	„ 26	2	„ 14	„ 8	
	*Great Spearwort (*Ranunculus lingua*), fl.	„ 26	6	„ 14	„ 10	
	French Willow-herb (*Epilobium angustifolium*), fl.	„ 26	:	...	...	Jun. 26
27	Cuckoo (*Cuculus canorus*), last heard	„ 27	10	Jun. 6	July 11	
	White-plume Moth (*Pterophorus pentadactylus*), ap.	„ 27				
	Couch-grass (*Triticum repens*), fl.	„ 27				
	Amphibious Nasturtium (*Nasturtium amphibium*), fl.	„ 27	3	Jun. 7	July 15	Jun. 28
	*Privet (*Ligustrum vulgare*), fl.	„ 27	7	„ 9	„ 20	July 9
	Yellow Toadflax (*Linaria vulgaris*), fl.	„ 27	5	„ 16	„ 8	
	Squinancy-wort (*Asperula cynanchica*), fl.	„ 27	4	„ 20	„ 7	
	Meadow Soft-grass (*Holcus lanatus*), fl.	„ 27	2	„ 25	Jun. 29	
	Cherries ripe	„ 27				
28	Wood Warbler (*Sylvia sibilatrix*), sg. ceas.	„ 28	2	Jun. 15	July 11	
	Wasp-Beetle (*Clytus arietis*), ap.	„ 28	2	„ 14	„ 12	
	Hairy St. John's Wort (*Hypericum hirsutum*), fl.	„ 28	5	„ 14	„ 7	Jun. 30
	*Field Scabious (*Knautia arvensis*), fl.	„ 28	8	„ 14	„ 19	July 6
	*Great Plantain (*Plantago major*), fl.	„ 28	5	„ 17	„ 8	
	Hemlock (*Conium maculatum*), fl.	„ 28	3	„ 17	„ 19	July 22

28	Round-leaved Bell-flower (*Campanula rotundifolia*), fl.	Jun. 28	7	May 27	July 27	July 19
29	Water Chickweed (*Malachium aquaticum*), fl.	,, 29	4	Jun. 8	,, 16	
	Flax-leaved Bastard-Toad-flax (*Thesium linophyllum*), fl.	,, 29	5	,, 18	,, 10	July 9
	†Great Knapweed (*Centaurea scabiosa*), fl.	,, 29	5	,, 20	,, 17	
	Whinchat (*Saxicola rubetra*), sg. ceas.	,, 30	3	,, 16	,, 10	
	Great Horse-fly (*Tabanus bovinus*), ap.	,, 30	6	,, 17	,, 16	July 1
	*Millefoil (*Achillea millefolium*), fl.	,, 30	7	,, 1	,, 18	
	*White Water-lily (*Nymphæa alba*), fl.	,, 30	4	,, 2	Aug. 2	
	White water Bedstraw (*Galium palustre*), fl.	,, 30	5	,, 8	July 14	July 26
30	Field Larkspur (*Delphinium consolida*), fl.	,, 30	6	,, 14	,, 25	,, 1
	†Meadow-sweet (*Spiræa ulmaria*), fl.	,, 30	2	,, 20	,, 15	
	Basil-Thyme (*Calamintha acinos*), fl.	,, 30	6	,, 20	,, 10	July 7
	†Common Agrimony (*Agrimonia eupatoria*), fl.	,, 30	6	,, 20	,, 20	
	Brook-weed (*Samolus valerandi*), fl.	,, 30	3	,, 28	,, 5	
	Meadow Pepper-saxifrage (*Silaus pratensis*), fl.	,, 30				

JULY.

JULY						
1	*Omaloptia ruricola*, ap.	July 1	2	Jun. 26	July 5	July 7
	Orange Lily (*Lilium bulbiferum*), fl.	,, 1		...	....	
	Water-dropwort (*Œnanthe fistulosa*), fl.	,, 1	3	Jun. 7	July 17	July 17
	†Blackberry (*Rubus fruticosus*), fl.	,, 1	8	,, 14	,, 15	,, 5
	*Barley (*Hordeum vulgare*), fl.	,, 1	5	,, 17	,, 13	
	Borage (*Borago officinalis*), fl.	,, 1	3	,, 20	,, 11	
	Pyramidal Orchis (*Orchis pyramidalis*), fl.	,, 1	5	,, 22	,, 9	Jun. 28
	Welted Thistle (*Carduus acanthoides*), fl.	,, 1	6	,, 23	,, 13	July 3
	*Spurge-laurel (*Daphne laureola*), berries ripe.	,, 1		...	...	

JULY	PHENOMENA	Mean.		Earliest.	Latest.	Date in 1845.
		(Date of occurrence; 1820—31.)				
	†Rooks return to their nest trees to roost (a)	July 2	6	Jun. 25	July 15	
	Second broods of House-sparrows hatched	,, 2				
	Rose Campion (*Lychnis coronaria*), fl.	,, 2	5			
	Common Skull-cap (*Scutellaria galericulata*), fl.	,, 2	3	Jun. 14	July 19	
2	Great wild Valerian (*Valeriana officinalis*), fl.	,, 2	4	,, 14	,, 24	
	Rest-harrow (*Ononis arvensis*), fl.	,, 2	4	,, 15	,, 10	
	Smooth-leaved Willow-herb (*Epilobium montanum*), fl.	,, 2	7	,, 16	,, 25	
	*Lime (*Tilia europæa*), fl.	,, 2	3	,, 21	,, 15	July 1
	*Red currants ripe	,, 2	3	,, 19	,, 9	,, 8
	Young Frogs come on land	,, 3	5	,, 28	,, 8	
	*Great Water-plantain (*Alisma plantago*), fl.	,, 3	4	,, 12	,, 23	July 26
3	Tufted Vetch (*Vicia cracca*), fl.	,, 3	5	,, 15	,, 20	
	Corn Sowthistle (*Sonchus arvensis*), fl.	,, 3	6	,, 21	,, 18	
	†Ragwort (*Senecio jacobæa*), fl.	,, 3	3	,, 24	,, 9	July 6
	*Raspberries ripe	,, 3	3	,, 14	,, 13	,, 5
	Young Jays (*Garrulus glandarius*) fledged	,, 4				
	Wood-leopard Moth (*Zeuzera æsculi*), ap.	,, 4	4	Jun. 18	July 20	
	Ghost Moth (*Hepialus humuli*), ap.	,, 4	3	,, 22	,, 11	
4	Dark Mullein (*Verbascum nigrum*), fl.	,, 4	4	,, 25	,, 10	
	Fiddle Dock (*Rumex pulcher*), fl.	,, 4	2	,, 26	,, 13	
	Sulphur-coloured Trefoil (*Trifolium ochroleucum*), fl.	,, 4	4	,, 27	,, 6	July 6
	*Common St. John's-wort (*Hypericum perforatum*), fl.	,, 4	6	,, 27	,, 17	
5	Chaffinch (*Fringilla cælebs*), sg. ceas.	,, 5	9			,, 4

	Phenomenon					
5	Tree Pipit (*Anthus arboreus*), lays a second time	July 5				
	Hen-harrier (*Circus cyaneus*), hatches	,, 5				
	Allecula sulphurea, ap.	,, 5				
	Silver-studded-blue Butterfly (*Polyommatus argus*), ap.	,, 5				
	Scarlet-tiger Moth (*Hypercompa dominula*), ap.	,, 5				
	*Cultivated Oat (*Avena sativa*), fl.	,, 5	4	Jun. 13	July 24	July 10
	*White Lily (*Lilium candidum*), fl.	,, 5	5	,, 14	,, 21	,, 6
	Flowering-rush (*Butomus umbellatus*), fl.	,, 5	6	,, 15	,, 19	
	Money-wort (*Lysimachia nummularia*), fl.	,, 5	5	,, 20	,, 17	
	Burnet-saxifrage (*Pimpinella saxifraga*), fl.	,, 5	2	,, 21	,, 20	
	Sharp Dock (*Rumex conglomeratus*), fl.	,, 5	2			
6	Young broods of Partridges fledged	,, 6				
	Young broods of Yellow-hammers fledged	,, 6				
	*Midsummer-Dor (*Melolontha solstitialis*), ap.	,, 6	2	Jun. 17	July 26	,, 19
	Ringlet Butterfly (*Hipparchia hyperanthus*), ap.	,, 6	6	,, 22	,, 17	
	Hairy wood Brome-grass (*Bromus asper*), fl.	,, 6	6	,, 17	,, 20	
	Yellow Bedstraw (*Galium verum*), fl.	,, 6	6	,, 22	July 18	
7	Lesser Whitethroat (*Curruca sylviella*), sg. ceas.	,, 7	7	,, 26	,, 30	
	†Glow-worm (*Lampyris noctiluca*), shines	,, 7	9	,, 25	,, 19	July 5
	Dwarf Thistle (*Carduus acaulis*), fl.	,, 7	7	,, 21	,, 17	
	*Spiked Speedwell (*Veronica spicata*), fl.	,, 7	6	,, 26	,, 18	July 1
	†White Jasmine (*Jasminum officinale*), fl.	,, 7	2	,, 29	,, 15	July 3
	Heath false Brome-grass (*Brachypodium pinnatum*), fl.	,, 7	7	,, 29	,, 16	
	Marsh Helleborine (*Epipactis palustris*), fl.	,, 7	4			
	*Gooseberries ripe	,, 7	3			July 23
8	Young broods of the common Lizard (*Zootoca vivipara*), ap.[1]	,, 8	3	Jun. 28	July 16	

(¹) They begin to return about this period; but the numbers keep gradually increasing as the summer advances.

JULY	PHENOMENA.	Date of occurrence; 1820—31.				Dat in 1845.
		Mean.		Earliest.	Latest.	
8	Lappet Moth (*Gastropacha quercifolia*), ap.	July 8	3	Jun. 13	July 23	July 9
	†Everlasting-pea (*Lathyrus latifolius*), fl.	„ 8	8	„ 28	„ 20	„ 8
	*Great Bindweed (*Convolvulus sepium*), fl.	„ 8	5	July 1	„ 17	„ 8
	Creeping Thistle (*Carduus arvensis*), fl.	„ 8	2	„ 6	„ 11	
	Slender false Brome-grass (*Brachypodium sylvaticum*), fl.	„ 8				
	Reed Meadow-grass (*Glyceria aquatica*), fl.	„ 8				
	Marjoram (*Origanum vulgare*), fl.	„ 8				
9	Tree Pipit (*Anthus arboreus*), sg., ceas.	„ 9	9	Jun. 20	July 20	July 21
	Shore Beetle (*Necrodes littoralis*), ap.	„ 9	2	July 1	„ 17	
	Autumnal Oporinia (*Oporinia autumnalis*), fl.	„ 9	5	Jun. 25	Aug. 1	July 23
	Hawkweed Picris (*Picris hieracoides*), fl.	„ 9	5	July 5	July 19	
	†Trees make their Midsummer shoots (a).	„ 9	5	Jun. 29	„ 27	
	Song Thrush (*Turdus musicus*), lays a second time	„ 10				
	Young broods of spotted Flycatchers fledged	„ 10				
10	Yellow-underwing Moth (*Triphæna pronuba*), ap.	„ 10	3	Jun. 25	July 18	July 20
	†Wild Succory (*Cichorium intybus*), fl.	„ 10	6	„ 15	„ 25	July 18
	Branched Bur-reed (*Sparganium ramosum*), fl.	„ 10	5	„ 27	Aug. 3	
	*Cat-mint (*Nepeta cataria*), fl.	„ 10	9	July 2	„ 2	
	Water-soldier (*Stratiotes aloides*), fl.	„ 10				
	Young Kestrels fledged (b)	„ 11				
11	Elephant Hawkmoth (*Deilephila elpenor*), ap.	„ 11	3	Jun. 24	July 29	
	Spotted Persicaria (*Polygonum persicaria*), fl.	„ 11	3	„ 18	Aug. 1	
	*Purple Loosestrife (*Lythrum salicaria*), fl.	„ 11	8	„ 20	„ 2	July 25

	Species					
11	Great Mullein (*Verbascum thapsus*), fl.	July 11	6	Jun. 26	Aug. 12	July 21
	Wild Basil (*Calamintha clinopodium*), fl.	,, 11	6	July 7	July 22	
12	Pettychaps (*Curruca hortensis*), sg., ceas.	,, 12	8	Jun. 28	,, 20	
	Magpie Moth (*Abraxas grossulariata*), ap.	,, 12	6	,, 29	,, 22	
	Silver-washed-fritillary Butterfly (*Argynnis paphia*), ap.	,, 12				
	Buckwheat (*Fagopyrum esculentum*), fl.	,, 12	4	Jun. 17	Aug. 1	
	Nettle-leaved Bell-flower (*Campanula trachelium*), fl.	,, 12	3	,, 21	July 31	
	Small-flowered Willow-herb (*Epilobium parviflorum*), fl.	,, 12	7	,, 28	,, 27	
	Great water Dock (*Rumex hydrolapathum*), fl.	,, 12	3	July 4	,, 19	
	Marsh Ragwort (*Senecio aquaticus*), fl.	,, 12	3	,, 4	,, 21	
	Traveller's-joy (*Clematis vitalba*), fl.	,, 12	2	,, 8	,, 16	
	†Houseleek (*Sempervivum tectorum*), fl.	,, 12				
13	*Hoplia argentea*, ap.	,, 13	3	Jun. 24	July 31	
	Fine-leaved Water-dropwort (*Œnanthe phellandrium*), fl.	,, 13	2	July 10	,, 16	
14	Drinker Moth (*Odanestis potatoria*), ap.	,, 14	4	Jun. 30	,, 26	July 27
	Marsh Woundwort (*Stachys palustris*), fl.	,, 14	6	July 5	Aug. 1	
	*Clove Pink (*Dianthus caryophyllus*), fl.	,, 14	3	,, 6	July 23	Aug. 1
	Common Centaury (*Erythræa pulchella*), fl.	,, 14	3	,, 10	,, 20	
15	White Horehound (*Marrubium vulgare*), fl.	,, 15	5	Jun. 7	Aug. 18	
	Black Nightshade (*Solanum nigrum*), fl.	,, 15	4	,, 17	,, 1	
	*Vervain (*Verbena officinalis*), fl.	,, 15	6	,, 17	,, 4	July 17
	Upright Hedge-parsley (*Torilis anthriscus*), fl.	,, 15	4	July 1	July 31	
	†Large-flowered St. John's-wort (*Hypericum calycinum*), fl.	,, 15	4	,, 3	,, 28	
	White Goose-foot (*Chenopodium album*), fl.	,, 15	4	,, 1	Aug. 7	July 25

(a) This is only a general observation, and perhaps of not much value. In repeating it, it would be well to mention in each case the particular tree.

(b) Probably second broods, as it will be seen by a former part of this calendar that the Kestrel lays April 22. I do not find mention, however, in authors, of this species breeding twice in the season.

July	Phenomena.	Date of occurrence; 1820—31.				Date in 1845.
		Mean.		Earliest.	Latest.	
15	Common Hemp-nettle (*Galeopsis tetrahit*), fl.	July 15	7	July 8	July 29	July 15
	*Peach-leaved Bell-flower (*Campanula persicifolia*), fl.	,, 15	10	Jun. 29	July 28	July 18
16	Blackbird (*Turdus merula*), sg. ceas.	,, 16	6	July 2	Aug. 1	,, 26
	†Large-heath Butterfly (*Hipparchia tithonus*), ap.	,, 16	2	July 2	July 30	
	Dark-green-fritillary Butterfly (*Argynnis aglaia*), ap.	,, 16	7	Jun. 28	,, 30	July 10
	V Moth (*Grammatophora vauaria*), ap.	,, 16	4	,, 30	,, 30	
	†Great Willow-herb (*Epilobium hirsutum*), fl.	,, 16	3	July 6	,, 25	
	White Poppy (*Papaver somniferum*), fl.	,, 16	2	,, 6	,, 25	
	Broad-leaved Water-parsnep (*Sium latifolium*), fl.	,, 16				
	Procumbent Water-parsnep (*Helosciadium nodiflorum*), fl.	,, 17				
17	Burnished-brass Moth (*Plusia chrysitis*), ap.	,, 17				
	Eyebright (*Euphrasia officinalis*), fl.	,, 17	5	Jun. 18	Aug. 12	
	Simple Bur-reed (*Sparganium simplex*), fl.	,, 17	4	,, 27	,, 10	
	Narrow-leaved Water-parsnep (*Sium angustifolium*), fl.	,, 17	4	July 1	,, 9	Aug. 4
	†Tutsan (*Hypericum androsaemum*), fl.	,, 18	5	,, 6	July 26	
	Pale-flowered Persicaria (*Polygonum lapathifolium*), fl.	,, 18	3	,, 10	Aug. 1	July 25
	Whitethroat (*Curruca cinerea*), sg. ceas.	,, 18	8	Jun. 28	July 28	
18	Wren (*Troglodytes europaeus*), second broods fledged		:	...	...	
	Dark-arches Moth (*Xylophasia polyodon*), ap.	,, 18	5	July 1	Aug. 12	July 18
	Cotton-thistle (*Onopordum acanthium*), fl.	,, 18				
	Spreading Hedge-parsley (*Torilis infesta*), fl.	,, 18	5	July 1	Aug. 12	
	*Blue Funkia (*Funkia ovata*), fl. { (*Hemerocallis caerulea*, Andr.) }	,, 18	:	...	...	July 25

	Phenomenon					
	Song Thrush (*Turdus musicus*), sg. ceas.	July 19	8	Jun. 26	July 27	July 18
	Reed Bunting lays a second time	,, 19	4	Jun. 29	Aug. 6	
19	Humming-bird Hawkmoth (*Macroglossa stellatarum*), ap.	,, 19	8	July 5	,, 12	
	Chalk-hill-blue Butterfly (*Polyommatus corydon*), ap.	,, 19	3	,, 10	July 23	
	Barred-lackey Moth (*Clisiocampa neustria*), ap.	,, 19	4	,, 8	Aug. 1	
	Great Reed-mace (*Typha latifolia*), fl.	,, 19	7	,, 9	,, 4	
	Square-stalked St John's-wort (*Hypericum quadrangulum*), fl.	,, 19	4	,, 10	,, 2	July 18
	Spear Thistle (*Carduus lanceolatus*), fl.	,, 20	2	,, 15	July 25	
	Goat Moth (*Cossus ligniperda*), ap.	,, 20	3	,, 15	,, 31	
	Garden-tiger Moth (*Arctia caja*), ap.	,, 20	9	Jun. 27	Aug. 7	July 8
	Small Teasel (*Dipsacus pilosus*), fl.	,, 20	4	,, 28	,, 3	
20	*Arrowhead (*Sagittaria sagittifolia*), fl.	,, 20	7	,, 28	,, 10	July 26
	Red Eyebright (*Euphrasia odontites*), fl.	,, 20	5	July 10	July 28	
	Corn Feverfew (*Pyrethrum inodorum*), fl.	,, 20	5	,, 12	,, 27	
	Fool's-parsley (*Ethusa cynapium*), fl.	,, 20	2	,, 17	,, 24	
21	Tit Pipit (*Anthus pratensis*), sg. ceas.	,, 21	4	,, 16	,, 27	
	Musk Beetle (*Cerambyx moschatus*), ap.	,, 21				
	Sedge Warbler (*Salicaria phragmitis*), sg. ceas.	,, 22	10	Jun. 28	Aug. 30	
22	*Necrophorus vestigator*, ap.	,, 22				
	†Hemp-agrimony (*Eupatorium cannabinum*), fl.	,, 22	8	July 9	Aug. 14	July 24
	Great-yellow Loosestrife (*Lysimachia vulgaris*), fl.	,, 22	5	,, 11	,, 5	
23	Willow Warbler (*Sylvia trochilus*), sg. ceas.	,, 23	9	Jun. 26	Sept. 13	July 4
	Turtle-dove (*Columba turtur*), last heard.	,, 23	5	July 18	Aug. 1	,, 26
	Swallow-tail Moth (*Ourapteryx sambucaria*), ap.	,, 24	2	,, 17	July 31	
24	†Flea-bane (*Pulicaria dysenterica*), fl.	,, 24	11	,, 2	Aug. 8	Aug. 1
	Burdock (*Arctium lappa*), fl.	,, 24	9	,, 9	,, 13	,, 1

JULY	PHENOMENA.	Date of occurrence; 1820—31.				Date in 1845.
		Mean.		Earliest.	Latest.	
25	Second broods of Goldfinches fledged	July 25				
	*Great-green Acrida (*Acrida viridissima*), stridulous note heard	,, 25	5	July 12	Aug. 9	
	Large-marsh Grasshopper (*Locusta flavipes*), ap..........	,, 25	5	July 2	Aug. 19	
	Common Star-thistle (*Centaurea calcitrapa*), fl	,, 25	2	,, 18	,, 1	
	*Apricots ripe	,, 26	9	,, 14	,, 23	
26	Hedge Accentor (*Accentor modularis*), sg. ceas.	,, 26	3	,, 12	,, 7	
	Grayling Butterfly (*Hipparchia semele*), ap..........	,, 26	2	,, 19	,, 2	
	*Dwarf Elder (*Sambucus ebulus*), fl	,, 26	10	Jun. 29	,, 18	July 25
27	Blackcap (*Curruca atricapilla*), sg. ceas.	,, 27	8	July 8	,, 13	Aug. 12
	†Wild Teasel (*Dipsacus sylvestris*), fl	,, 27	7	,, 11	,, 14	,, 12
	†Hairy Mint (*Mentha sativa*), fl.	,, 27	5	,, 15	,, 8	July 15
	Red Hemp-nettle (*Galeopsis ladanum*), fl	,, 27	4	Jun. 30	Aug. 11	
28	Young broods of Swifts (*Cypselus apus*), fledged	,, 28				
	†Admiral Butterfly (*Vanessa atalanta*), ap.........	,, 28	3	July 18	,, 11	
	Small-skipper Butterfly (*Pamphila linea*), ap.........	,, 28	3	,, 20	,, 6	
	Common Feverfew (*Pyrethrum parthenium*), fl	,, 28	5	,, 15	,, 9	
	Bastard Stone-parsley (*Sison amomum*), fl	,, 28	2	,, 27	July 29	
	Knotted Spurrey (*Spergula nodosa*), fl............	,, 28	2	,, 19	Aug. 9	
29	*Common Grasshopper (*Locusta biguttula*), crinks	,, 29	3	,, 12	,, 17	July 28
	Common Calamint (*Calamintha officinalis*), fl	,, 29	3	,, 25	,, 11	
	Fennel (*Fœniculum officinale*), fl.	,, 29				
30	Gipsey-wort (*Lycopus europæus*), fl............	,, 30	4	,, 25	,, 5	July 29

Day	Phenomenon					
30	*Wheat cut ...	12	July 30	July 16	Aug. 18	Aug. 18
31	Hoary Ragwort (*Senecio erucifolius*), fl.	5	,, 31	,, 19	,, 15	,, 8
31	Wild Angelica (*Angelica sylvestris*), fl.	4	,, 31	,, 25	,, 11	

AUGUST.

Day	Phenomenon					
1	Second broods of Swallows fledged		Aug. 1			
1	Wormwood (*Artemisia absinthium*), fl.	3	,, 1	July 24	Aug. 11	
2	Large-eggar Moth (*Lasiocampa quercus*), ap.	1	,, 2			
2	Large-flowered Hemp-nettle (*Galeopsis versicolor*), fl.	2	,, 2			
4	Mugwort (*Artemisia vulgaris*), fl.	4	,, 4	July 26	Aug. 8	Aug. 6
5	Mushrooms (*Agaricus campestris*), abound	3	,, 5	Jun. 20	Sept. 20	
6	*Honey-suckle berries ripe		,, 6			
8	Cole Titmouse (*Parus ater*), note ceas.	3	,, 8	July 13	Sept. 22	Aug. 8
8	*Swift (*Cypselus apus*), last seen	8	,, 8	,, 29	Aug. 23	
	Hornet-fly (*Asilus crabroniformis*), ap.	.	...			
8	Carline-thistle (*Carlina vulgaris*), fl.	3	,, 8	Aug. 6	Aug. 10	
9	Common Linnet (*Linota cannabina*), sg. ceas.	8	,, 9	July 18	,, 30	
9	Purple Melic-grass (*Molinia caerulea*), fl.	2	,, 9	,, 27	,, 22	
9	Red Goose-foot (*Chenopodium rubrum*), fl.		,, 9			
10	Silver-spotted-skipper Butterfly (*Pamphila comma*), ap	9	,, 10			
12	Yellow-hammer (*Emberiza citrinella*), sg. ceas.		,, 12	July 29	Aug. 27	
12	Ringdove lays a second time [a]		,, 12			
12	*Swallows and Martins begin to congregate	8	,, 12	July 23	Aug. 25	
12	Soapwort (*Saponaria officinalis*), fl.	3	,, 12	,, 25	Sept. 1	
13	Second broods of house Martins fledged		,, 13			Aug. 13

[a] Perhaps the third, instead of the second time.

AUG.	PHENOMENA.	Mean.		Earliest.	Latest.	Date in 1845.
13	Zabrus gibbus, ap.	Aug. 13	4	July 21	Sept. 11	
14	Goldfinch (Carduelis elegans), sg. ceas.	,, 14	5	Aug. 3	Aug. 31	
15	Large-black Staphyline (Goerius olens), ap.	,, 15	5	July 6	Sept. 15	
16	Greenfinch (Coccothraustes chloris), sg. ceas.	,, 16	9	,, 24	,, 8	
16	*Artichoke (Cynara scolymus), fl.	,, 16				
17	*Barley cut	,, 17	4	Aug. 1	Aug. 30	
18	*Devil's-bit Scabious (Scabiosa succisa), fl.	,, 18	5	,, 5	,, 31	
18	*Bracts of the Lime fall (a)	,, 18	3	July 29	Sept. 8	
19	Common Tansy (Tanacetum vulgare), fl.	,, 19				
20	Redbreast (Erithaca rubecula), sg. reass.	,, 20	5	Aug. 7	Sept. 10	Aug. 12
20	Small-copper Butterfly (Lycæna phlæas), ap.	,, 20	2	,, 17	Aug. 23	Aug. 23
20	Woolly-headed Thistle (Carduus eriophorus), fl.	,, 20	3	,, 12	Sept. 2	
23	Gold-spot Moth (Plusia festucæ), ap.	,, 23				Aug. 30
24	*Starlings collect in flocks	,, 24	9	July 12	Oct. 9	,, 24
25	*Winged Ants migrate	,, 25	9	,, 26	Sept. 28	
29	Martins collect in great numbers on the roofs of houses	,, 29				
30	*Red Bryony berries ripe.	,, 30	2	Aug. 8	Sept. 21	
31	Autumnal Gentian (Gentiana amarella), fl.	,, 31	4	,, 12	,, 8	
31	*Peaches ripe	,, 31	2	,, 24	,, 7	

SEPTEMBER.

SEPT.						
1	Clouded-yellow Butterfly (*Colias edusa*), ap.	Sept. 1	2	Aug. 23	Sept. 11	
2	Chaffinch (*Fringilla cœlebs*), sg. reass.	,, 2	5	,, 22	,, 19	
	*Barberries ripe	,, 2	2	,, 24	,, 11	
3	*Meadow-saffron (*Colchicum autumnale*), fl.	,, 3				
5	Stock-dove (*Columba œnas*), note ceas.	,, 5				
	*Hawthorn berries ripe	,, 5	3	Sept. 1	Sept. 8	Sept. 20
6	†Gossamer floats	,, 6	9	Aug. 18	,, 23	
	Swallow (*Hirundo rustica*), sg. ceas.	,, 7	7	,, 11	Oct. 2	
7	*Red-underwing Moth (*Catocala nupta*), ap.	,, 7	4	,, 11	,, 1	Sept. 12
	*Cuckow-pint (*Arum maculatum*), berries ripe.	,, 7	3	,, 24	Sept. 17	Aug. 30
9	Dog-rose casts its leaves	,, 9				
10	Great Titmouse (*Parus major*), sg. reass.	,, 10	3	Sept. 6	Sept. 17	Sept. 25
	*Yew-berries ripe	,, 10	2	,, 2	,, 18	
12	House-flies swarm in windows	,, 12	3	,, 5	,, 17	
13	*Elder-berries ripe	,, 13	3	,, 12	,, 14	Oct. 1
15	House Sparrows collect in large flocks	,, 15				
	Vapourer Moth (*Orgyia antiqua*), ap.	,, 15	2	Sept. 8	Sept. 23	Sept. 29
	Lime turns yellow.........	,, 16	7	Aug. 28	,, 23	Oct. 1
16	Horse-chestnut turns brown	,, 16	2	Sept. 12	,, 20	
	Wich Elm leaves begin to fall	,, 16	3	,, 7	,, 27	
	Gold-crested Wren (*Regulus cristatus*), sg. reass.	,, 18	5	,, 9	Oct. 1	Sept. 9
18	Peewits collect in flocks	,, 18	3	Aug. 25	,, 5	
	Goldfinches collect in flocks	,, 18	5	Sept. 5	,, 29	Sept. 8
	Sycamore leaves fall	,, 18	2	,, 15	Sept. 22	Oct. 12

(ᵃ) This is synchronous with the ripening of the fruit, to which the bracts are attached.

SEPT.	PHENOMENA.	Date of occurrence; 1820—31.				Date in 1846.
		Mean.		Earliest.	Latest.	
19	Dotterel makes its autumnal passage	Sept. 19	3	Sept. 17	Sept. 22	Sept. 22
	Acorns fall	,, 19	4			
20	Chiffchaff (*Sylvia hippolais*), note ceas.	,, 20	5	Sept. 9	Sept. 29	Sept. 29
	Syringa (*Philadelphus coronarius*), turns yellow	,, 20	7	,, 2	Oct. 8	Oct. 3
21	†Drone-fly (*Eristalis tenax*), enters houses	,, 21	7	Aug. 24	,, 15	Sept. 27
	Tutsan (*Hypericum androsæmum*), turns brown	,, 21				
22	Lime leaves begin to fall	,, 22	7	Sept. 2	Oct. 7	Oct. 2
23	Herald Moth (*Calyptra libatrix*), ap.	,, 23	5	Aug. 26	Nov. 9	
24	Beech-mast falls	,, 24	4	Sept. 4	Oct. 22	Oct. 4
26	Hedge Accentor (*Accentor modularis*), sg. reass.	,, 26	4	,, 14	,, 6	Sept. 30
	*Ivy (*Hedera helix*), fl.	,, 26	11	,, 5	,, 22	Nov. 9
27	Birch turns yellow	,, 27	4	,, 21	,, 10	
	Ringdove (*Columba palumbus*), note ceas.	,, 28	7	,, 14	,, 9	
28	*Acrida varia*, ap.	,, 28	4	,, 14	,, 30	
	Laurestine (*Viburnum tinus*), fl.	,, 28	5	Aug. 19	Dec. 16	
	*Martins (*Hirundo urbica*), bulk departed	,, 29	.		...	Oct. 30
29	Autumn-green-carpet Moth (*Harpalyce miata*), ap.	,, 29	2	Sept. 25	Oct. 3	,, 10
	Beech turns yellow	,, 29	9	,, 6	,, 23	

OCTOBER.

OCT.	PHENOMENA.	Mean.		Earliest.	Latest.	Date in 1846.
1	Common Snipe (*Scolopax gallinago*), ap. in plenty (ᵉ)	Oct. 1	7	Aug. 20	Nov. 14	
	Maple turns yellow	,, 1	1			

No.	Phenomenon					
1	Horse-chestnut leaves begin to fall	Oct. 1	5	Aug. 27	Oct. 14	
2	Walnut leaves begin to fall	,, 2	6	Sept. 20	Oct. 28	Oct. 8
3	Jack-Snipe (*Scolopax gallinula*), arrives	,, 3	8	,, 6	Nov. 13	,, 5
	*Linnets (*Linota cannabina*), arrives ... collect in flocks	,, 4	3	,, 5	,, 8	
4	Sloes ripe	,, 4	4	…	…	
	Horse-chestnuts fall	,, 4	4	…	…	
	*Walnuts ripe	…	…	…	…	
5	Virginian-Creeper (*Ampelopsis hederacea*), turns red	,, 5	6	Sept. 13	Oct. 28	Oct. 2
	Trees in general assume their autumnal tints	,, 5	10	,, 22	,, 22	
	Birch leaves fall	,, 5	3	,, 30	,, 10	
6	Buntings (*Emberiza miliaria*), collect in flocks	,, 6	6			
7	Wheat sown	,, 7	7	Sept. 23	Oct. 19	Oct. 6
	Maple leaves fall	,, 7	7			
	Beech leaves begin to fall	,, 7	7			
8	White Poplar leaves begin to fall	,, 8	8	Sept. 25	Oct. 19	Oct. 12
	Cherry leaves fall	,, 8	8			
9	Hazel turns yellow	,, 9	9	Oct. 5	Oct. 14	Oct. 5
11	Elm turns yellow	,, 11	11	,, 4	,, 18	,, 18
	Ash leaves begin to fall (b)	,, 11	11	Sept. 20	,, 28	,, 12
12	Honeysuckle leaves fall	…	…	…	…	
	Crab-apples ripe and falling	…	…	…	…	,, 12
13	Aspen leaves fall	,, 13	2	Oct. 6	Oct. 20	
	Lombardy Poplar leaves fall	,, 13	3	,, 11	,, 15	
	Elder leaves fall	,, 13	3			

(a) Many Snipes remain with us the whole summer, but there is an accession of numbers in the autumn.

(b) Linnæus observes of the Ash, "*Fraxinus inter primas defoliatur et inter ultimas frondescit.*" (Phil. Bot.) This is often the case. The defoliation, however, depends a good deal upon the time of occurrence of the first sharp frost, which frequently brings all the leaves down together, so that a tree is nearly stript in a night. The consequence is, that the leaves fall *green*, unlike those of any other tree.

T

Oct.	PHENOMENA.	Mean.		Earliest.	Latest.	Date in 1845.
14	*Swallow (*Hirundo rustica*), last seen........	Oct. 14	12	Sept. 28	Oct. 31	Oct. 15
	Ladybird (*Coccinella septempunctata*), hybernates	,, 14	2	Oct. 8	,, 20	Oct. 15
16	*Martin (*Hirundo urbica*), last seen.	,, 16	11	,, 2	Nov. 4	
17	Dogwood turns red	,, 17				
	Hazel leaves begin to fall	,, 17				
18	*Lime stript of its leaves	,, 18	7	Sept. 22		Oct. 18
19	Yellow-hammer (*Emberiza citrinella*), sg. reass. .	,, 19	3	Oct. 14		
	Virginian-Creeper leaves fall	,, 19	4	,, 16		
21	*Walnut stript of its leaves	,, 21	2	,, 5		Nov. 3
22	Coddy-moddy Gull (*Larus canus*), comes inland....	,, 22				
24	Golden Plover (*Charadrius pluvialis*), arrives ..	,, 24	2	Sept. 22	Nov. 25	
25	†Short-eared Owl (*Otus brachyotos*), arrives	,, 25	3	,, 21	,, 15	
26	Whitethorn leaves fall	,, 26				
28	*Flocks of Wild-Geese (*Anser segetum*), arrive	,, 28	12	Oct. 7	Nov. 29	
	*Wild-Duck (*Anas boschas*), arrives	,, 29	3	,, 7	,, 21	
29	Virginian-Creeper stript	,, 29	4	,, 25	,, 1	Oct. 29
	*Red Currant stript	...	:	...	...	,, 29
	*Syringa stript	...	:	...	...	,, 27
30	*Woodcock (*Scolopax rusticola*), arrives	,, 30	9	,, 18	,, 18	Nov. 2
	*White Poplar stript......	,, 30	:	...	...	
31	Elm leaves begin to fall	,, 31	4	Oct. 18	Nov. 15	Oct. 16

NOVEMBER.

Nov.						
1	*Sycamore stript of leaves					Nov. 1
	*Hazel stript					,, 1
	*Guelder-rose stript	Nov. 2				,, 1
2	Missel Thrush (*Turdus viscivorus*), sg. reass	,, 2	5	Sept. 20	Dec. 11	Oct. 27
	*Horse-chestnut stript	,, 3	3	Oct. 19	Nov. 15	Nov. 3
3	Plane leaves fall		3	Oct. 27	Nov. 11	Oct. 28
	*Lilac stript	Nov. 4				Nov. 4
	*Ash stript					,, 4
	Apple-tree stript					,, 4
4	*Gooseberry stript					,, 4
	*Laburnum stript					,, 4
	*Hornbeam stript					
	*Whitethorn stript					Oct. 13
5	Skylark (*Alauda arvensis*), sg. ceas	Nov. 5	4	Oct. 9	Dec. 10	
	*Cherry-tree stript	,, 5				Nov. 8
7	*Hooded Crow (*Corvus cornix*), arrives	,, 7	3	Oct. 9	Dec. 5	,, 8
	Larch turns yellow	,, 8				
8	Lombardy Poplar stript					Nov. 15
	*Birch stript	Nov. 10				
10	Bunting (*Emberiza miliaria*), note ceas	,, 13	2	Oct. 11	Dec. 9	
13	†Woodpigeons collect in flocks	,, 15	8	,, 11	,, 20	
	Larch leaves fall					
15	*Apricot stript	,, 15				Oct. 28
	*Beech stript	,, 16				

Nov.	PHENOMENA.	Date of occurrence; 1820—31.			Date in 1845.
		Mean.		Earliest.	Latest.	

Nov.	PHENOMENA.	Mean.		Earliest.	Latest.	Date in 1845.
16 {	Titmice draw near houses.........	Nov. 16	8	Oct. 17	Dec. 17	Nov. 1
	Teal (*Anas crecca*), arrives	,, 16	2	,, 3	,, 31	
21 {	*Fieldfare (*Turdus pilaris*), arrives	,, 21	9	,, 26	,, 21	
	Redwing (*T. iliacus*), arrives	,, 21	3	,, 29	,, 21	
24 {	Grey Wagtail (*Motacilla boarula*), arrives	,, 24	3	,, 10	Jan. 8	
	Larch stript.........	,, 24		....	....	Nov. 4
26	*Oak stript.......	,, 26	.	....	....	,, 16
28	*Elm stript...	,, 28	2	Nov. 26	Nov. 30	,, 15
29	Song Thrush (*Turdus musicus*), sg. reass.....	,, 29	6	,, 15	Dec. 13	

DECEMBER.

Dec.	PHENOMENA.	Mean.		Earliest.	Latest.	Date in 1845.
1	Trees everywhere stript of leaves......	Dec. 1	6	Nov. 27	Dec. 12	Nov. 20
2	*Pipistrelle Bat last seen abroad....	,, 2	8	Oct. 25	,, 31	
8	Skylarks collect in flocks.......	,, 8	8	Nov. 26	,, 13	
15	Greenfinches collect in flocks.....	,, 15	2	Dec. 13	,, 17	
25 {	Chaffinches collect in flocks......	,, 25				
	Marsh Titmouse (*Parus palustris*), sg. reass....	,, 25				

ALPHABETICAL ARRANGEMENT

OF THE

PERIODIC PHENOMENA IN THE FOREGOING CALENDAR,

WITH A REFERENCE TO THE MEAN DATE OF OCCURRENCE.

N.B. Only the English names are given in general.

A.

Accentor, hedge, song com-
         mences ...... Jan. 17
  „ hatches .... Apr. 18
  „ young broods fledged „ 23
  „ song ceases........ July 26
  „ song reassumed .... Sept. 26
Aconite, winter, flowers . Jan. 26
Acorns, fall............ Sept. 19
Acrida, great green, stridu-
        lous note heard July 25
Acrida varia, ap. ...... Sept. 28
Agrimony, common, fl. .. June 30
Alder, fl. .... Feb. 28
  „ leafs .... Apr. 11
Alkanet, evergreen, fl. .. Apr. 9
Allecula sulphurea, ap. .. July 5
Almond, fl............ Mar. 28
American-Cowslip, fl. .. Apr. 26
Anemony, snowdrop, fl... May 13
  „ wood, fl.......... Apr. 10
Angelica, wild, fl. ...... July 31
Ant, red, ap. .... Mar. 15
Ants, winged, migrate .. Aug. 25

Apple-tree, stript ...... Nov. 4
Apricot, fl............ Mar. 10
  „ leafs .... Apr. 3
  „ stript .... Nov. 15
Apricots ripe .... July 25
Archangel, yellow, fl..... May 13
Arrow-head, fl. .... July 20
Artichoke, fl........... Aug. 16
Ash, fl. .... Apr. 11
  „ leafs .... „ 30
  „ leaves begin to fall.. Oct. 11
  „ stript .... Nov. 4
Aspen, fl............. Mar. 17
  „ leafs .... May 11
  „ leaves fall .... Oct. 13
Atopa cervina, ap. ...... June 19
Avens, water, fl........ May 14

B.

Badister bipustulatus, ap. Mar. 22
Barberries, ripe .... Sept. 2
Barberry, l. .... Apr. 3
  „ flowers .... „ 29
Barley, sown .... Mar. 21

Brome-grass, slender, false, fl. July 8

„ soft, fl. . . . . May 21

„ upright, fl. . . . . June 13

Brook-lime, fl. . . . . May 27

Brook-weed, fl. . . . . June 30

Bryony, black, fl. . . . . . . June 10

„ red, fl. . . . . May 31

„ berries ripe . . . . Aug. 30

Buckbean, fl. . . . . May 18

Buckthorn, fl. . . . . June 7

„ alder, fl. . . . . May 17

Buckwheat, fl. . . . . July 12

Bugle, fl. . . . . May 3

Bugloss, Italian, fl. . . . . . . June 5

„ small, fl. . . . . „ 11

Bull-rush, fl. . . . . „ 26

Bunting, common, note commences . . . . Jan. 11

„ note ceases . . . . Nov. 10

Bunting, reed, song commences . . . . . . Apr. 28

„ lays. . . . . May 11

„ lays a second time . . July 19

Buntings collect in flocks Oct. 6

Burdock, fl. . . . . July 24

Burnet, lesser, fl. . . . . . . May 11

Burnet-saxifrage, fl. . . . . July 5

Bur-reed, branched, fl. . . „ 10

„ simple, fl. . . . . „ 17

Butcher's-broom, fl. . . . . Feb. 12

Butter-bur, fl. . . . . Apr. 24

Buttercup, fl. . . . . „ 25

Butterfly, admiral, ap. . . July 28

„ brimstone, ap. . . . . . . Mar. 10

„ brown-argus, ap. . . . . . May 26

„ chalk-hill-blue, ap. . . July 19

„ clouded-yellow, ap. . . . Sept. 1

„ common-blue, ap. . . June 5

„ dark-green-fritillary, ap. July 16

Butterfly, dingy-skipper, ap. May 14

„ grayling, ap. . . . . . . July 26

„ greasy-fritillary, ap. May 23

„ green-veined-white, ap. Apr. 26

„ grizzled-skipper . . . . May 26

„ large-heath, ap. . . . . July 16

„ large-skipper, ap. . . June 14

„ large-tortoise-shell, ap. Apr. 29

„ large-white, ap. . . . . „ 23

„ meadow-brown, ap. . . June 19

„ orange-tip, ap. . . . . . . May 8

„ painted-lady, ap. . . . . June 24

„ peacock, ap. . . . . Mar. 31

„ ringlet, ap. . . . . July 6

„ silver-spotted-skipper, ap. Aug. 10

„ silver-studded-blue, ap. July 5

„ silver-washed-fritillary, ap. „ 12

„ small-blue, ap. . . . . June 13

„ small-copper, ap. . . Aug. 20

„ small-heath, ap. . . . . May 24

„ small-skipper, ap. . . July 28

„ small-tortoise-shell, ap. Mar. 27

„ small-white, ap. . . . . Apr. 11

„ speckled-wood, ap. . . „ 30

„ swallow-tail, ap. . . . . May 30

„ wall, ap. . . . . „ 12

Butter-wort, fl. . . . . „ 25

C.

Campion, rose, fl. . . . . . . July 2

„ white, fl. . . . . May 22

Calamint, common, fl. . . July 29

Carabus nemoralis, ap. . . Mar. 24

Carex, great, fl. . . . . May 8

Carline-thistle, fl. ...... Aug. 8
Carrot, wild, fl. .... June 20
Cat-mint, fl. .... July 10
Cat's-ear, long-rooted, fl. June 14
„ spotted, fl. .... „ 26
Celandine, fl. .... May 11
Centaury, common, fl..... July 14
Chaffer, small garden, ap. June 5
Chaffinch, song commences Feb. 1
„ builds .... Apr. 14
„ lays .... „ 28
„ hatches .... May 4
„ young broods fledged „ 17
„ song ceases .... July 5
„ song reassumed .... Sept. 2
Chaffinches collect in flocks Dec. 25
Chamomile, wild, fl. .... June 16
Charlock, fl. .... May 11
Cherries, ripe.......... June 27
Cherry, l. .... Apr. 3
„ flowers .... „ 14
„ leaves fall .... Oct. 8
„ stript .... Nov. 5
Chervil, rough, fl. ...... June 4
„ wild, fl. .... Apr. 18
Chickweed, field, fl...... May 4
„ water, fl........... June 29
Chiffchaff, note first heard Apr. 3
„ note ceases .... Sept. 30
Cinque-foil, creeping, fl... June 16
„ strawberry-leaved, fl. Apr. 15
Clary, English, fl. ...... May 13
Clover, red, fl. .... „ 7
„ white, fl........... „ 22
Cockchaffer, ap. .... „ 15
Cocksfoot-grass, fl. ...... June 7
Coltsfoot, common, fl..... Mar. 13
„ sweet-scented, fl. .. Feb. 23
Columbine, fl.......... May 13
Comfrey, fl............ „ 12

Cord-moss, hygrometric, ripens its capsules Mar. 18
Corn-cockle, fl. .... June 18
Corn-sallad, fl. .... May 13
Cotton-thistle, fl. ...... July 18
Couch-grass, fl. .... June 27
Cow-parsnep, fl......... „ 4
Cowslip, fl............. Mar. 30
Crab, l. .... Apr 3
„ flowers .... „ 27
Crab-apples, ripe and falling Oct. 12
Crane's-bill, bloody, fl. .. June 11
„ dove's-foot, fl...... May 3
„ jagged-leaved, fl.... June 8
„ meadow, fl........ „ 17
Creeper, spring note commences ...... Mar. 11
„ builds .... May 6
„ lays .... „ 7
Creophilus maxillosus, ap. June 5
Crocus, spring, fl. ...... Feb. 19
Crosswort, fl........... May 20
Crow, hooded, last seen.. Mar. 9
„ arrives .... Nov. 7
Crowfoot, celery-leaved, fl. May 16
„ corn, fl. .... May 30
„ creeping, fl........ Mar. 22
„ upright, fl. .... May 16
„ water, fl........... Apr. 24
„ wood, fl........... „ 21
Crown-imperial, fl....... „ 5
Cuckoo, first heard...... „ 27
„ last heard .... June 27
Cuckow-pint, fl. .... Apr. 30
„ berries, ripe........ Sep. 7
Cudweed, common, fl. .. June 21
„ mountain, fl. ...... „ 7
Currant, black, l. ...... Mar. 19
„ flowers .... Apr. 17
Currant, red, l. .... Mar. 24

Currant, red, flowers.... Apr. 11
„ stript .... Oct. 29
Currants, red, ripe ...... July 2
Cushion-moss, grey, ripens
    its capsules.... Mar. 5

D.

Daffodil, fl............. Mar. 13
Daisy, fl. .... Jan. 28
„ double, fl. .... „ 30
Dandelion, fl........... Feb. 18
Daphne, trailing, fl. .... Apr. 14
Day-lily, tawny, fl. .... June 25
„ yellow, fl. .... May 29
Dead-nettle, henbit, fl. .. Apr. 26
„ red, fl. .... Feb. 15
„ white, fl. .... Mar. 5
Deadly-Nightshade, fl. .. June 5
Dew-berry, fl. .... May 25
Dock, broad-leaved, fl. .. June 22
„ curled, fl. .... „ 12
„ fiddle, fl........... July 4
„ great water, fl. .... „ 12
„ sharp, fl........... „ 5
Dog-rose, l. .... Mar. 18
„ flowers .... June 9
„ casts its leaves .... Sept. 9
Dog-rose, trailing, fl..... June 16
Dog-wood, l. .... Apr. 24
„ flowers .... June 9
„ turns red .... Oct. 17
Dogs-tail-grass, fl. ...... June 23
Dor, midsummer, ap. .... July 6
Dor-beetle, ap. .... Mar. 6
Dotterel, makes its spring
    passage ...... Apr. 27
„ makes its autumnal
    passage ...... Sept. 19
Dragon-fly, four-spotted, ap. May 31
„ golden-green, ap... . „ 21

Dragon-fly, large-brown, ap. June 10
Drone-fly, enters houses . Sept. 21
Drop-wort, fl........... June 25
Duck, domestic, hatches.. Mar. 31
Dung of cattle, swarms
    with coleopterous insects Apr. 30

E.

Earthworms, lie out .... Feb. 23
Earwig, ap. .... Mar. 25
Eft, small smooth, ap. in
    ponds .... Feb. 8
Elater lineatus, ap....... Mar. 21
„ *sputator,* ap. ...... „ 22
Elder, common, l. ...... Feb. 14
„ flowers .... June 1
„ berries ripe .... Sept. 13
„ leaves fall .... Oct. 13
Elder, dwarf, fl......... July 26
Elm, common, fl. ...... Mar. 21
„ leafs .... Apr. 7
„ sheds its seed ...... May 5
„ turns yellow ...... Oct. 11
„ leaves begin to fall .. Oct. 31
„ stript .... Nov. 28
Elm wych, fl........... Apr. 1
„ leafs .... „ 29
„ leaves begin to fall .. Sept. 16
Empis pennipes, ap. .... May 5
Enchanter's-nightshade, fl. June 23
Everlasting-pea, fl. .... July 8
Eye-bright, fl. .... „ 17
„ red, fl............. „ 20

F.

Fennel, fl. .... July 29
Feverfew, common, fl. .. „ 28
„ corn, fl. .... „ 20
Fieldfare, arrives ...... Nov. 21
Fieldfares, last seen .... Apr. 5

Guelder-rose, flowers .... May 26
,, stript .... Nov. 1
Gull, coddy-moddy, comes
     inland ...... Oct. 22
Gymnostomum ovatum, ripens
     its capsules.... Feb. 9

H.

Hair-grass, crested, fl. .. June 25
Harpalus æneus, ap. .... Mar. 27
,, *ruficornis,* ap. .... May 16
Harry-long-legs, ap. .... ,, 10
Hawkbit, rough, fl. .... June 14
Hawk-moth, elephant, ap. July 11
,, eyed, ap. .... June 20
,, humming-bird, ap... July 19
,, lime, ap. .... Apr. 30
,, poplar, ap. ...... ,, 27
,, privet, ap. .... June 26
,, small elephant, ap. ... ,, 12
Hawks-beard, smooth, fl. { June 21
                      { ,, 22
Hawk-weed, mouse-ear, fl. May 24
Hawthorn-berries, ripe .. Sept. 5
Hay, meadow, cut...... June 11
Hazel, fl. .... Jan. 25
,, leafs .... Mar. 29
,, turns yellow ...... Oct. 9
,, leaves begin to fall.. " 17
,, stript .... Nov. 1
Heart's-ease, fl. .... Apr. 11
Hedge-mustard, common, fl. May 14
Hedge-parsley, spreading,
     flowers ...... July 18
,, upright, fl. .... ,, 15
Hellebore, stinking, fl. .. Jan. 28
Helleborine, marsh, fl. .. July 7
Hemlock, fl. .... June 28
Hemp-agrimony, fl...... July 22
Hemp-nettle, common, fl.. ,, 15

Hemp-nettle, large flow-
     ered, fl. .... Aug. 2
,, red, fl............ July 27
Henbane, fl............ June 15
Henharrier, hatches .... July 5
Hepatica, fl. .... Jan. 25
Herb-bennet, fl. .... May 23
Herb-Robert, fl......... ,, 4
Hister unicolor, ap. .... ,, 1
Holly, fl. .... ,, 19
Honey-suckle, common, l. Feb. 18
,, flowers .... June 8
,, berries ripe ...... Aug. 6
,, leaves fall .... Oct. 12
Honey-suckle, pale perfo-
     liate, l. ...... Jan. 15
,, flowers .... May 6
Hoplia argentea, ap. .... July 13
Horehound, stinking, fl... June 23
,, white, fl.......... July 15
Hornbeam, l........... Apr. 15
,, flowers .... ,, 21
,, stript .... Nov. 4
Hornet-fly, ap. .... Aug. 8
Horse-chestnut, l. ...... Mar. 31
,, flowers .... May 3
,, turns brown ...... Sept. 16
,, leaves begin to fall.. Oct. 1
,, stript .... Nov. 2
Horse-chestnuts fall .... Oct. 4
Horse-fly, great, ap. .... June 30
,, small, ap. .... ,, 25
Horse-shoe-vetch, tufted, fl. May 26
Horse-tail, corn, fl...... Apr. 19
Hound's-tongue, fl. .... May 28
House-flies appear in win-
     dows .... Jan. 27
" swarm in windows.. Sept. 12
Houseleek, fl.......... July 12
Humble-bee ap........ Mar. 20

Humble-bee-fly, ap. .... Apr. 11
Hyacinth, fl. .... Mar. 27

I.

Insects, broods of small cole-
     opterous, on wing Mar. 7
Iris, dwarf purple, fl. .... Mar. 29
 ,, german, fl. .... Apr. 25
Ivy, berries ripe........ Apr. 1
 ,, casts its leaves .... June 16
 ,, flowers .... Sept. 26

J.

Jack-by-the-hedge, fl. .. Apr. 23
Jackdaw builds .... Mar. 25
 ,, lays .... Apr. 23
Jackdaws resort to chim-
     neys .... Feb. 18
 ,, young, fledged .... June 11
Jacob's-ladder, fl. ...... May 25
Jasmine, white, l. ...... May 9
 ,, flowers .... July 7
Jays, young, fledged .... ,, 4

K.

Kerria, japan, l. .... Feb. 24
 ,, flowers .... Mar. 15
Kestrel lays .... Apr. 22
 ,, young broods fledged July 11
Knapweed, black, fl. .... June 15
 ,, great, fl........... ,, 29

L.

Laburnum, l. .... Apr. 7
 ,, flowers .... May 11
 ,, stript .... Nov. 4
Lady-bird, ap. .... Mar. 4
 ,, hybernates........ Oct. 14
Lady's-fingers, fl. ...... June 14
Lady's-smock, meadow, fl. Apr. 18

Landrail, note first heard June 6
 ,, lays .... ,, 14
Larch l. .... Apr. 2
 ,, flowers .... ,, 8
 ,, turns yellow ...... Nov. 8
 ,, leaves fall .... ,, 15
 ,, stript .... ,, 24
Larkspur, bee, fl........ June 25
 ,, field, fl. .... ,, 30
Laurel, common, fl. .... Apr. 10
 ,, Portugal, fl. ...... { June 24 / ,, 25
Laurestine, fl. .... Sept. 28
Libellula depressa, ap. ... May 30
Lilac, common, l. ...... Mar. 18
 ,, flowers .... Apr. 30
 ,, stript .... Nov. 3
Lilac, Persian, fl. ...... May 7
Lily, orange, fl. .... July 1
 ,, turk's-cap, fl. ...... June 18
 ,, white, fl........... July 5
Lily-of-the-valley, fl..... May 12
Lime l. .... Apr. 12
 ,, flowers .... July 2
 ,, bracts fall .... Aug. 18
 ,, turns yellow ...... Sept. 16
 ,, leaves fall .... ,, 22
 ,, stript .... Oct. 18
Linnet, common, song com-
     mences .... Mar. 23
 ,, lays .... Apr. 25
 ,, young broods fledged May 28
 ,, song ceases........ Aug. 9
Linnets collect in flocks .. Oct. 4
Lizard, common, ap. .... Apr. 11
 ,, young broods ap.. July 8
London-pride, fl. ...... Apr. 30
Loosestrife, great yellow, fl. July 22
 ,, purple, fl. .... ,, 11
Lousewort, marsh, fl. .. May 30

Lungwort, Virginian, fl.   Apr. 2

M.

Magpie builds ....   Mar. 22
Malachius æneus ap. ....   May 21
Mallow, common, fl. ....   June 11
„   dwarf, fl. ....   „ 11
Maple, l. ....   Apr. 15
„   flowers ....   „ 29
„   turns yellow ......   Oct. 1
„   leaves fall ....   „ 7
Mare's-tail, fl. ....   June 15
Marjoram, fl. ....   July 8
Marsh-marigold, fl. ......   Mar. 4
Martin, house, first seen   Apr. 30
„   builds ....   May 21
„   second broods fledged Aug. 13
„   last seen ....   Oct. 16
Martin, sand, first seen ..   May 7
Martins collect in great
    numbers on the
    roofs of houses ...   Aug. 29
„   bulk departed .....   Sept. 29
May-fly ap. ....   May 17
May-weed, fl. ....   June 24
Meadow-grass, flat stalked,
    flowers ......   May 31
„   floating, fl. ....   June 14
„   reed, fl. ....   July 8
„   rough stalked, fl....   June 12
„   smooth stalked, fl...   May 28
Meadow-rue, fl. ....   June 14
Meadow-saffron, fl. ....   Sept. 3
Meadow-sweet, fl. ......   June 30
Mealy-tree, l. ....   Mar. 31
„   flowers ....   May 2
Medick, black, fl. ......   May 3
Medlar, fl. ....   „ 3
Melic-grass, purple, fl. ..   Aug. 9
Melilot, fl. ....   June 15

Mercury, dog's, fl. ......   Mar. 26
Mezereon, fl. ....   Jan. 22
Midge, ap. ....   May 18
Mignonette, wild, fl. ....   May 22
Milk-vetch, purple-moun-
    tain, fl. ......   „ 23
„   sweet, fl. ....   June 20
Milkwort, fl. ....   May 12
Millefoil, fl. ....   June 30
Mint, hairy, fl. ....   July 27
Misseltoe, fl. ....   Mar. 4
Mite, scarlet satin, ap. ..   „ 17
Money-wort, fl. ....   July 5
Monk's-hood, fl. ....   May 18
Moorhen lays ....   Apr. 6
„   hatches ....   „ 23
Morell, ap. ....   „ 23
Moth, autumn-green-car-
    pet, ap. ......   Sep. 29
„   barred-lackey, ap. ..   July 19
„   bright-line brown-
        eye, ap. ......   June 10
„   buff-tip, ap. ....   May 10
„   burnished-brass, ap.   July 17
„   dagger, ap. ....   June 9
„   dark-arches, ap. ....   July 18
„   dot, ap. ....   May 14
„   drinker, ap. ....   July 14
„   emperor, ap. ......   May 8
„   garden-carpet, ap. ..   May 27
„   garden-tiger, ap.....   July 20
„   ghost, ap. ....   „ 4
„   goat, ap. ....   „ 20
„   gold-spot, ap. ......   Aug. 23
„   heath, ap. ....   May 24
„   herald, ap. ....   Sept. 23
„   lappet, ap. ....   July 8
„   large-eggar, ap. ....   Aug. 2
„   latticed-heath, ap. ..   May 5
„   magpie, ap. ....   July 12

Moth, mullein, ap....... May 4
  „ pale-tussock, ap. .... „ 10
  „ pink-underwing, ap. June 14
  „ puss, ap........... May 31
  „ red-underwing, ap... Sept. 7
  „ scarlet-tiger, ap. .... July 5˙
  „ silver-Y, ap. ...... June 10
  „ six-cleft-plume, ap... Mar. 27
  „ six-spot-burnet, ap. June 22
  „ swallow-tail, ap..... July 24
  „ tissue, ap. .... Apr. 25
  „ V, ap............. July 16
  ,, vapourer, ap. ...... Sept. 15
  „ white-plume, ap. .. June 27
  „ wood-leopard, ap. .. July 4
  ,, yellow-underwing,
        appears .... July 10
Mountain-ash, fl. ...... May 7
Mouse-ear-chickweed, nar-
        row-leaved, fl. Apr. 21
Mugwort, fl. .... Aug. 4
Mulberry, l. .... May 19
  ,, flowers .... June 10
Mullein, dark, fl. ...... July 4
  ,, great, fl........... „ 11
Mushrooms abound .... Aug. 5
Mustard, white, fl...... June 11

N.

Narcissus, pale, fl. .... Apr. 22
Nasturtium, amphibious,
        flowers ...... June 27
Navel-wort, blue, fl. .... Mar. 19
Necrophorus vestigator,
        appears .... July 22
Nettle, great, fl......... June 6
  „ small, fl........... May 31
Nightingale first heard .. Apr. 21
  ,, lays .... May 13
  „ song ceases........ June 7

Nightshade, black, fl..... July 15
  ,, woody, fl. ...... June 13
Nipplewort, fl. .... „ 22
Nostoc, jelly, appears on
        lawns........ Apr. 22
Nuthatch, whistling note
        heard........ „ 26

O.

Oak l. .... May 2
  ,, flowers .... „ 8
  ,, stript .... Nov. 26
Oak, evergreen, fl. ...... May 27
Oat, cultivated, fl....... July 5
  ,, wild, fl. .... June 24
Oat-grass, fl. .... June 15
  „ downy, fl. .... „ 4
  ,, narrow-leaved, fl... June 16
Oats, sown............ Mar. 17
Oiceoptoma rugosa, ap. .. Apr. 26
  „ thoracica, ap. ...... „ 16
Omaloplia ruricola, ap. .. July 1
Oporinia, autumnal, fl. .. July 9
Orchis, bee, fl. .... June 19
  „ butterfly, fl. ...... „ 10
  „ dwarf-dark-winged,
        flowers ...... May 24
  ,, early purple, fl. .... „ 26
  ,, fly, fl............. „ 14
  „ green-winged mea-
        dow, fl. ...... „ 12
  „ marsh, fl........... „ 31
  „ pyramidal, fl....... July 1
  „ spotted-palmate, fl.. June 6
  „ sweet-smelling, fl. .. „ 21
Owl, long-eared, lays.... Apr. 16
  ,, short-eared, arrives.. Oct. 25
  „ tawny, hoots ...... Feb. 2
  „ ,, lays ..... Mar. 28
Ox-eye-daisy, fl........ May 27

P.

Pæony, fl. .... May 15
Pancratium, sea, fl....... May 23
Parsnep, wild, fl. ...... June 10
Partridge, pairs .... Feb. 13
„ lays .... May 10
Partridges, young, hatched June 21
„ young, fledged .... July 6
Pasque-flower, fl. ...... Apr. 20
Peach, fl............... Mar. 13
„ leafs .... Apr. 2
Peaches, ripe .... Aug. 31
Peacock, screams ...... Mar. 14
Pear, leafs .... Apr. 8
„ flowers .... „ 11
Peewit, lays .... Apr. 19
Peewits, collect in flocks Sept. 18
Pellitory-of-the-wall, fl... June 19
Pepper-saxifrage, meadow,
     flowers ...... June 30
Periwinkle, lesser, fl..... Feb. 22
Persicaria, amphibious, fl. July 20
„ pale-flowered, fl. .. „ 17
„ spotted, fl. .... „ 11
Pettychaps, first heard .. May 4
„ song ceases........ July 12
Peziza coccinea, ap. .... Jan. 31
Pheasant, utters its spring
     crow . ... Apr. 3
„ lays .... May 6
Pheasants, young, hatched June 6
Picris, hawkweed, fl..... July 9
Pigeon, house, lays...... Feb. 11
„ „ hatches .... Mar. 8
Pilewort, fl. .... „ 2
Pimpernel, scarlet, fl..... June 8
Pink, clove, fl. .... July 14
Pipit, tit, song commences Apr. 11
„ lays .... May 14
„ song ceases........ July 21

Pipit, tree, first heard .. Apr. 20
„ lays .... May 18
„ lays a second time.. July 5
„ song ceases........ „ 9
Plane, l. .... May 6
„ leaves fall .... Nov. 3
Plantain, great, fl....... June 28
„ hoary, fl. .... May 27
„ ribwort, fl......... Apr. 27
Plover, golden, arrives .. Oct. 24
„ great, first heard or
     seen .... Apr. 17
Plum, l................. „ 4
„ flowers .... „ 7
Pæcilus cupreus, ap. .... Mar. 26
Poplar, black, fl. ...... Apr. 4
„ leafs .... „ 23
Poplar, Lombardy, fl. .. „ 4
„ leafs .... „ 23
„ leaves fall .... Oct. 13
„ stript .... Nov. 8
Poplar, white, fl. ...... Mar. 25
„ leafs .... Apr. 19
„ leaves fall .... Oct. 8
„ stript .... „ 30
Poppy, common red, fl. .. June 3
„ large oriental, fl..... „
„ long-prickly-headed,
     flowers ...... „ 6
„ white, fl. .... July 16
Primrose, fl, ... Feb. 11
Privet, l, ... Mar. 28
„ flowers .... June 27
Pyrochroa rubens, ap..... „ 4

Q.

Quail, note first heard .. May 29
Quaking-grass, fl. ...... June 15
Quince, fl. .... Apr. 26

R.

Radish, wild, fl......... May 24
Ragged-robin, fl. ...... „ 19
Ragwort, fl. .... July 3
„ hoary, fl. .... „ 31
„ marsh, fl. .... „ 12
Raspberries, ripe ...... July 3
Raspberry, l. .... Apr. 8
„ flowers .... May 22
Rattle, yellow, fl. ...... „ 30
Redbreast, song commences Jan. 8
„ lays .... Apr. 11
„ hatches .... „ 14
„ young fledged .... „ 29
„ lays a second time .. June 4
„ second broods hatched „ 12
„ song ceases........ „ 23
„ song reassumed .... Aug. 20
Redstart, first heard .... Apr. 15
„ song ceases........ June 13
„ young broods fledged „ 15
Redwing, arrives ...... Nov. 21
Reedmace, great, fl. .... July 19
Rest-harrow, fl. .... July 2
Rhingia rostrata, ap. .... May 2
Rhyphus fenestralis, ap. .. Mar. 12
Ringdove, coos .... Feb. 25
„ lays .... Apr. 25
„ hatches .... „ 30
„ lays a second time .. Aug. 12
„ note ceases........ Sept. 28
Rocket, dyer's, fl. ...... June 17
Rock-rose, common, fl. .. May 26
Rooks build .... Mar. 3
„ lay .... Mar. 25
„ hatch .... Apr. 9
„ young fledged .... May 2
„ return to their nest-
trees to roost .. July 2
Rose, burnet, l......... Apr. 10

Rose, burnet, fl......... June 4
„ moss, fl. .... „ 14
Rush, hard, fl. .... „ 10
Rye, fl. .... „ 5
Rye-grass, fl........... „ 4

S.

Saint-foin, fl. .... June 7
Sand-piper, common, first
seen .... May 28
Sand-wort, thyme-leaved,
flowers ...... June 3
Sanicle, wood, fl. ...... May 28
Saxifrage, rue-leaved, fl... Apr. 6
Scabious, devil's-bit, fl. .. Aug. 18
„ field, fl. .... June 28
„ small, fl........... „ 20
Scorpion-fly, ap......... May 13
Scorpion-grass, field, fl. .. May 18
„ wood, fl........... „ 27
Self-heal, fl. .... June 20
Shepherd's-needle, fl..... May 10
Sherardia, blue, fl....... „ 14
Sialis lutarius, ap. ...... Apr. 25
Silpha obscura, ap....... „ 16
„ *lævigata*, ap. ...... „ 18
Silver-weed, fl. .... May 18
Skullcap, common, fl..... July 2
Skylark, song commences. Jan. 23
„ song ceases........ Nov. 5
Skylarks, collect in flocks. Dec. 8
Sloes, ripe .... Oct. 4
Slug, black, ap. .... { Apr. 16
                       { „ 17
Snail, banded, comes a-
broad .... Mar. 20
„ „ engenders .. Apr. 29
„ common, comes abroad „ 16
„ „ engenders .. „ 26
„ heath, comes abroad. Mar. 1

Snake, ringed, comes a-
  broad .... Apr. 8
  „   „ couples .... „ 22
Snap-dragon, great, fl. .. June 11
Snipe, common, appears in
  plenty .... Oct. 1
  „ jack, arrives ...... „ 3
Snowdrop, fl. .... Jan. 30
Soap-wort, fl. .... Aug. 12
Soft-grass, meadow, fl. .. June 27
Solomon's-seal, fl. ...... May 15
Sorrel, common, fl. ...... „ 28
Sow-thistle, fl. .... June 16
  „ corn, fl. .... July 3
Sparrow, house, builds .. Apr. 10
  „ second broods hatched July 2
Sparrows, house, collect in
  flocks .... Sept. 15
Spearwort, great, fl. .... June 26
  „ lesser, fl. .... „ 11
Speedwell, common, fl. ... „ 8
  „ field, fl. .... Feb. 4
  „ germander, fl. ...... Apr. 29
  „ ivy-leaved, fl. ...... Mar. 11
  „ spiked, fl. .... July 7
  „ thyme-leaved, fl. .. May 9
  „ water, fl, ... June 19
Spiderwort, Virginian, fl. .. June 2
Spike-rush, creeping, fl. ... „ 18
Spurge, caper, fl. ...... May 2
Spurge-laurel, fl. ...... Feb. 5
  „ berries ripe .... July 1
Spurrey, knotted, fl. .... July 28
Squinancy-wort, fl. .... June 27
Squirrel builds .... Apr. 23
St. John's-wort, common, fl. July 4
  „ hairy, fl. .... June 28
  „ large flowered, fl. .. July 15
  „ square-stalked, fl. .. „ 19
Staphylinus, large black, ap. Aug. 15

Star-of-Bethlehem, fl. .... May 15
Star-thistle, common, fl. ... July 25
Starlings, resort to build-
  ings .... Jan. 21
  „ young broods fledged May 19
  „ collect in flocks .... Aug. 24
Steropus madidus, ap .... Apr. 20
Stinging-fly, ap. .... May 29
Stock-dove, note com-
  mences ...... Mar. 5
  „ lays .... Apr. 13
  „ note ceases .... Sept. 5
Stone-crop, biting, fl. .... June 22
Stork's-bill, hemlock fl. ... May 11
Strawberries, ripe ...... June 22
Strawberry, fl. .... Apr. 26
Succory, wild, fl. ...... July 10
Swallow, first seen ...... Apr. 19
  „ song commences .... „ 23
  „ builds .... May 7
  „ young broods fledged June 14
  „ second broods fledged. Aug. 1
  „ song ceases .... Sept. 7
  „ last seen .... Oct. 14
Swallows and Martins
  congregate .... Aug. 12
Sweetbriar, l. .... Mar. 21
  „ flowers .... June 21
Sweet-William, fl. ...... „ 17
Swift, first seen .... May 13
  „ young broods fledged . July 28
  „ last seen .... Aug. 8
Sycamore, l. .... Apr. 12
  „ flowers .... „ 29
  „ leaves fall .... Sept. 18
  „ stript .... Nov. 1
Syringa, l. .... Mar. 17
  „ flowers .... May 31
  „ turns yellow ...... Sept. 20
  „ stript .... Oct. 29

T.

Tansy, common, fl. .... Aug. 19
Teal, arrives .... Nov. 16
Teasel, small, fl........ July 20
„ wild, fl. .... „ 27
Thistle, creeping, fl. .... July 8
„ dwarf, fl........... „ 7
„ marsh, fl.......... June 19
„ milk, fl. .... „ 17
„ musk, fl........... „ 12
„ spear, fl........... July 19
„ welted, fl. .... „ 1
„ woolly-headed, fl. .. Aug. 20
Thrincia, hairy, fl...... May 25
Thrush, missel, song com-
    mences ...... Jan. 26
„ lays .... Apr. 9
„ hatches .... „ 20
„ song ceases........ May 27
„ song reassumed .... Nov. 2
Thrush, song, song com-
    mences ...... Jan. 31
„ lays .... Apr. 12
„ hatches .... „ 20
„ young fledged .... „ 30
„ lays a second time.. July 10
„ song ceases ...... „ 19
„ song reassumed .... Nov. 29
Thyme, wild, fl........ June 11
Titmice,   draw   near
    houses ...... Nov. 16
Titmouse, cole, note com-
    mences ...... Jan. 19
„ note ceases........ Aug. 8
Titmouse, great, song com-
    mences ...... Jan. 24
„ song ceases........ May 22
„ song reassumed .... Sept. 10
Titmouse, long-tailed, lays Apr. 18
„ hatches .... May 12

Titmouse, marsh, note com-
    mences ...... Jan. 1
„ note ceases........ Apr. 18
„ song re-assumed.... Dec. 25
Toad, common, spawns.. Mar. 27
Toad-flax, yellow, fl .... June 27
Tormentil, common, fl. .. May 16
Traveller's-joy, fl. ...... July 12
Treacle-mustard, worm-
    seed, fl. .... June 15
Trees, make their mid-
    summer shoots . July 9
„ assume their autum-
    nal tints ...... Oct. 5
„ everywhere stript of
    leaves........ Dec. 1
Trefoil, hop, fl. .... June 24
„ strawberry, fl. .... June 26
„ sulphur-coloured, fl. July 4
Trichocera hiemalis, ap... Jan. 11
Tulip, wild, fl. .... Apr. 10
Turkey-cock struts and
    gobbles ...... Mar. 10
Turnip, fl.............. Apr. 9
Turnip-fly, ap. .... Apr. 6
Turtle-dove, first heard.. May 8
„ lays .... June 18
„ last heard .... July 23
Tutsan, fl. .... July 17
„ turns brown Sept.21
Tway-blade, fl. .... May 17

V.

Valerian,   great   wild,
    flowers ...... July 2
„ red, fl .... June 8
„ small marsh, fl. .... May 8
Velia currens appears on
    streams .... Feb. 21

GENERAL INDEX.

N

Phenomena, tabular arrangement of those most worthy of record, 341.

—— average range of variation of, 346.

—— the mean date of, how obtained, 347, *note*.

—— a selection of, best for purposes connected with climatological inquiries, 348.

—— M. Quetelet's plan for observing, 349, 351.

—— M. Quetelet's directions for observers of, 354.

Philodromus limacum, a mite found on slugs, some particulars respecting, 291.

Pike, instances of its voracity, 217.

—— mode of taking in the fens, 218.

Pike-louse, notice respecting, 296.

Pipistrelle, taken on wing, with the young adhering to the breast, 58.

Pipit, tree, nest and eggs of, 135.

Plover, great, notes respecting, 177.

—— ringed, numbers in the fens in the summer of 1824, 179.

Pontia cardamines, the males apparently more plentiful than the females, 264.

Prejudices, those likely to interfere with correct observation, 33, 35.

Ptinus fur, destructive to books, 239.

Q.

Quadrupeds, the mode in which various kinds suckle their young, 51.

Quail, remarks on its note, 176.

R.

Rabbits, sometimes found with the incisor teeth of monstrous growth, 78.

—— many found dead with their livers full of flukes, 80.

Rail, water, notes respecting, 182.

Raven, one that ate a bat entire, 140.

Razorbill, one picked up alive near Wimpole, 191.

Redbreast, anecdote of, 127.

Redbreast, instances of its pugnacious disposition, 129, 130.

—— singing in severe frost, 130.

Red-eye, or shallow, note respecting, 214.

Red-spider, a mite that infests fruit trees, some account of, 293.

Redwing, remarks on the nature of its food, 125.

Rookeries, strong smell proceeding from, 146.

Rooks, their different times of coming home to roost, 140.

—— often restless during the night, 142.

—— remain much upon their nest trees before the breeding-season actually commences, 143.

—— sound the note of alarm to their companions on the approach of danger, 143.

White's Natural History of Selborne, the influence it has had on the present generation, 10.

Whitethroat, lesser, its nest and eggs, 131.

Woodpecker, great-spotted, seen at Bottisham mostly in spring and autumn, 155.

—— much infested by a minute *Acarus,* 155.

Worm, a peculiar kind infesting the gills of smelts, 219.

—— one found in a gnat, 311.

Worms, remarks on those inhabiting the land and fresh-water, 298.

—— thread, instance of great numbers found after rain, 303.

THE END.

ERRATA.

Page 81, line 2 from the bottom, *for* calculi, *read* concretions.

„   89   „ 12 from the top, *for* hear, *read* heard.

„   298   „   6 from the bottom, *for* molluscks, *read* mollusks.

„   299   „   3 from the bottom (notes), the asterisk (*) should be a dagger (†).

LONDON:
Printed by S. & J. BENTLEY, WILSON, and FLEY,
Bangor House, Shoe Lane.

Printed in the United States
by Booksurge

Printed in the United States
By Bookmasters